AMERICA
THE
INGENIOUS

KEVIN BAKER

AMERICA
THE
INGENIOUS

HOW A NATION OF DREAMERS, IMMIGRANTS, AND TINKERERS CHANGED THE WORLD

Illustrations by Chris Dent

ARTISAN

NEW YORK

Library of Congress Cataloging-in-Publication Data

Names: Baker, Kevin, 1958– author.
Title: America the ingenius / Kevin Baker.
Description: New York, NY : Artisan, a division of Workman Publishing
    Company, Inc., [2016] | Includes bibliographical references.
Identifiers: LCCN 2016027529 | ISBN 9781579656942 (hardback, with dust jacket)
Subjects: LCSH: Inventions—United States—Popular works. | Inventors—United
    States—Popular works.
Classification: LCC T21 .B35 2016 | DDC 609.73–dc23 LC record available at
https://lccn.loc.gov/2016027529

Design by Jacob Covey | Unflown

Artisan books are available at special discounts when purchased in bulk for premiums and sales promotions as well as for fund-raising or educational use. Special editions or book excerpts also can be created to specification. For details, contact the Special Sales Director at the address below, or send an e-mail to specialmarkets@workman.com.

Published by Artisan
A division of Workman Publishing Company, Inc.
225 Varick Street
New York, NY 10014-4381
artisanbooks.com

Artisan is a registered trademark of Workman Publishing Co., Inc.

Published simultaneously in Canada by Thomas Allen & Son, Limited

Printed in the United States

First printing, October 2016

10 9 8 7 6 5 4 3 2 1

To all the niblings, Ari, Anik, Jackson, Zoe, Julian, and Griffin,
with high hopes for the wondrous world they will grow up in

And as always, with love
to Ellen

CONTENTS

APPAREL

WOMEN INVENTORS

BUILDING

POWERING

FIGHTING

CURING

PRODUCING

PLAYING

THE AMERICAN GENIUS

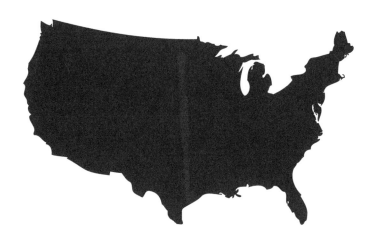

The American genius is a unique one. As befits a nation of immigrants, America has excelled at taking ideas from anywhere and transforming them into the practical, startling, often unexpectedly beautiful creations that have shaped our world.

We are a nation of tinkerers, never content to leave a thing alone. Born inventors, researchers, adventurers, we keep seeking what's next, what's better, what's more fun, what will make more money. As the first country to exist wholly in the modern age, we have also had to invent and constantly *re*invent ourselves—our institutions, our customs, even our definition of who is an American, and what that means.

These are the stories of some of the best things that we've taken up—who did them, how they were done, what they meant. The list is an eclectic one, and highly subjective, moving freely from the skyscraper to the subway, the safety elevator to the space elevator, the Pennsylvania rifle to the video game. They include many of our most important and famous innovations, such as the telephone, the transistor, and the farm combine, as well as more pedestrian inventions that nonetheless transformed daily life for millions: the sewing machine, air-conditioning, the electric guitar—all the way down to the humble safety pin.

Here you will find many things that aren't generally included in books about inventions: blues and jazz, our greatest cities, the Tennessee Valley Authority (TVA). Why not? They were invented by people just as much as anything else in this book, and they are constantly cited by others as some of our greatest creations.

Here, as well, are inventions whose utility ended years ago, but that changed the world in their day, such as the telegraph, or the Yankee clipper. Inventions that have declined in importance, such as dried blood plasma, but may yet again play a crucial role in the near future. Innovations that were simply, stunningly beautiful, such as the streamlined automobiles, planes, and trains of the 1930s—and which also represented a gigantic step forward in technological research and application.

And here, too, you will find a glimpse of an equally dazzling future that is already beginning to emerge. Will we zip around in flying cars—or driverless ones? Build prosthetics that work better than our arms and legs? Climb into space on a single cable the thickness of a thread? The answers are likely to be as amazing as all the earlier creations here first seemed when they came into being.

The inventions in this book are inextricably linked to the American experiment. They were inspired, shaped, and made possible by the character of our country.

This is not to say that Americans invented everything. Far from it. Throughout history, very little was invented solely in one nation or by one person working alone. The "American" inventions cited here were inspired by ideas, theories, and prototypes going all the way back to the toga-cleaning facilities of ancient Greece and Rome; the ninth-century printing presses of China, Korea, and Japan; even the iron-chain suspension bridges thrown across the chasms of the Himalayas by a Buddhist saint in 1430.

The Western world at the outset of the industrial age was a particular hotbed of invention. New discoveries and their applications poured out of Europe and, especially, Great Britain, and credit often remains disputed to this day. To peruse the Internet is to gain the idea that everything in the world was invented by the English, which is to say, a Scotsman.

The fact is that many of the inventions in this book were indeed first conceived of, or even first made a working reality, in other countries, and to the extent that space allows, they are freely credited here. Others were invented at virtually the same time in America as they were elsewhere, with the advances in each nation sometimes feeding off one another.

In every case, though, each one of these devices, structures, remedies, systems, styles, enterprises, and entertainments were fully realized in America. If they were not wholly invented in the United States, it was here that they were made commercially viable, widespread, affordable, beloved, indispensable.

More important than whether Americans came up with everything first is the fact that we were once, at least, able and willing to learn from others. To take what was dreamed up anywhere around the world and improve on it. (Or even, in the case of magnetic tape recording, cart the basic idea off as a spoil of war.)

What were some of the other keys to our inventiveness?

FREEDOM. Yes, *you* did build that. What comes across at every turn in studying the history of American innovation is that this is what free men (and women) can do when afforded the liberty to pursue whatever they wish. Whether it's the genius of an Alexander Graham Bell, the dogged persistence of a Thomas Edison, the entrepreneurship of an Andrew Carnegie or a Henry Ford, the courage of a Walter Reed, or even the sheer, exasperating eccentricity of a Walter Hunt, the value of the individual shines through.

WE ALL BUILT THAT. Beyond the obvious heroes, I have tried here to demonstrate how much we owe to all the "ordinary" people who made our progress a reality. The generations of anonymous pioneers who gave us the Conestoga wagon, and then adapted it into the prairie schooner. The trail of craftsmen along the eighteenth-century frontier who perfected and built the Pennsylvania rifle. The nameless Irish navvies who dug out the Erie Canal by hand; the black and white sandhogs who looked death in the face every day as they carved out the tunnels beneath New York's rivers.

Their contributions, and those of countless others, were as valuable as anyone's in making us all that we are and giving us all that we have.

GOVERNMENT MATTERS. Rugged individualism aside, the history of American invention shows again and again that government—which in a democracy, again, means all of us—is vital. It was government that set the rules, with patent law that settled what otherwise would have been interminable legal battles over inventions from the cotton gin to the automobile; that rescued countless small inventors, such as Philo Farnsworth, inventor of electronic television, and Elias Howe, of sewing machine fame, from predatory forces that would have stolen their creations; that set up antitrust regulations, and in so doing spurred the invention of everything from the modern oil rig to Silicon Valley; that so often put up the cash to sponsor projects from the Trancontinental Railroad to the space race, the telegraph to the Hoover Dam; that provided the money for a first-class public education system and countless research grants and laboratories; that has freely "picked winners," right down to Abe Lincoln testing a repeating rifle himself on the Washington Mall during the Civil War.

IMMIGRATION IS CRUCIAL. Over and over, in researching this book, I was struck by just how much of America was made by immigrants. Many of these, of course, were those anonymous men and women who did the hard work of hauling and digging, riveting and welding. But beyond arm and back, so many contributed their brains, as well. What would America have been without them? Not just the more famous ones such as Bell, Carnegie, or A. P. Giannini, but also Carl Breer, son of a German immigrant, who led a revolution in car design; or Richard Hoe, son of an English immigrant printer, who gave us the rotary press; or Jacob Youphes and Loeb Strauss, two young Jewish men who came to America, changed their names to Jacob Davis and Levi Strauss, and gave us blue jeans.

For many, many years, we have drawn the best and the brightest from nations all around the globe. America's lifeblood is immigration, and to ever shut it off would be fatal.

A MIND IS A TERRIBLE THING TO WASTE. Readers will find among our inventors a relatively small proportion of women and people of color. This is because both were prevented for many years from filing patents by law, and then by social custom and persecution.

How many enslaved African Americans were deprived of the fruits of their labor for things they actually invented? How many women saw their achievements purloined by husbands, fathers, bosses? We shall never know, but it is instructive to note that the number of patents won by women today is over seven times the number they earned a hundred years ago.

Even denied so many outlets for so very many years for their advancement and education, African Americans invented everything from a blood plasma system to a revolutionary advance in removing cataracts—not to mention what much of the world considers to be the very greatest of American accomplishments, *our* music, blues and jazz.

IT TAKES A VILLAGE. Again and again, bringing brilliant and talented people together produces magic.

This is true not just in a purely academic setting, though American colleges and universities have made myriad contributions to the commonwealth. From Edison's workshops and the Volta Laboratory down through Bell Labs, the NASA center at Huntsville, Alabama, and Silicon Valley, concentrations of all sorts of smart people have created vital nodes of creation and commerce. Proximity matters.

Perhaps the most amazing thing about American ingenuity is, well, how many different ways we've invented to invent things. Take the basic elements of individual initiative, collaboration, education, government funding, venture capitalism, immigration, innovation, hard work, obstinacy, obsession, persistence, irrational optimism, and the occasional earth-shaking epiphany, and mix them together in any number of different ways. It all works, providing us with a sort of built-in redundancy to our great experiment in democracy. And as we continue, I am sure that we will find still more ways to make it work.

In the meantime, here it is to explore, presented, I hope, in a clear and easily comprehensible manner. Presented, as well, on a human scale, and through the works, the hopes, and the dreams of the sort of regular human beings who together can do anything. This is their story. ★

THE PRAIRIE SCHOONER

They spread out across the prairie like the flag itself: a sea of wagons, with their canvas coverings painted red, white, or blue. There were so many of them that this was the only way they could keep track of who belonged to which wagon train. A nation on the move, setting off across a continent at the pace of an ox.

The American covered wagon was not invented by any one individual but was a tremendous piece of folk craft, perfected over 150 years by countless anonymous carpenters, village smiths, and pioneers.

The wagons that conquered the West evolved out of the Conestogas, first mentioned in accounts going back to 1717: long, heavy wagons with low, curved tops, suited to the rutted roads of Mennonite country in Pennsylvania's Lancaster County. With their bodies painted blue and their wheels a bright red, they looked more like European tinkers' wagons and were primarily vehicles for hauling freight. German and Scotch-Irish immigrants took them down the Great Wagon Road to the Shenandoah Valley and the Blue Ridge Mountains, then out to Ohio and Illinois.

It took the covered wagon four to six months to cross the West. Most wagons had wooden or leather hoops on the inside for hanging milk cans, clothes, guns, or even dolls.

The Conestogas were fine for these relatively short hauls, but something more was needed to take on the American West. By the time the pioneers started pushing out from St. Louis in force in the 1840s, the vehicles they used were lighter, higher, tougher, carefully calibrated for journeys of two thousand miles across deserts and mountains, rivers and swamps.

"The prairie schooner," as it came to be known, was about the size of a family minivan today. Sturdy yet maneuverable, it pivoted on an iron or steel kingpin connected to a wagon tongue and two massive axles. Its wheels were big and wide and rimmed with iron to keep them from getting bogged down in mud or soft earth. The wagon box was generally ten feet long by four feet wide, its hardwood "Yankee bed" waterproofed so the whole wagon could be floated across rivers.

The double-canvas top that gave the prairie schooner its name was tall enough for a man to stand up under it. Often made of hemp, it was stretched out over five or six high hickory bows and waterproofed with paint or linseed oil. The whole family might sleep under it in bad weather.

Covered wagons were pieces of extraordinary workmanship. Just welding on the two pieces of the wheel rims was a painstaking art that required making the iron hot enough to fit securely but not so hot as to burn the wooden wheel. They sold for $250 and up, and a "proper outfit" of wagon, supplies, and the ten to twelve horses, mules, or preferably oxen needed to haul them usually cost $800 to $1,000. Families often saved for three, or four, or five years to buy everything. They rarely rode in them. Built without springs, they jounced enough that pioneers could leave a bucket of milk in the wagon at first light and see it churned to butter by nightfall.

Those who could walk, did. Besides, the wagons weren't for the settlers but for their possessions. Prairie schooners weighed about 1,300 pounds themselves and could haul 3,000 pounds, but they rarely packed more than two-thirds of that weight. This might include bags of seed, an ax, a rifle, small pieces of furniture, bedding, clothes, and other finished goods that would be hard to come by on the frontier: a skillet, a Dutch oven, a coffeepot, an oil lamp, a spinning wheel—though such items often ended up lining one last mountain pass that was just too steep. Most of what they carried was food: flour, lard, and bacon wrapped in bran, to be augmented by whatever game or fish pioneers could take on the way, enough of it to get them where they were going and then through the year or so it took to plant and harvest a crop there.

The prairie schooner's back wheels were bigger, five to six feet in diameter, compared to just four feet or less for the front ones, which allowed the wagon to make sharp turns.

The wagon trains could usually cover fifteen to twenty miles a day at a speed of about two miles an hour. The pioneers left notes for each other, scribbled on anything they could find, warning of wrong turns or bad water through the "Prairie Post Office" or the "Roadside Telegraph." They died along the way, too, twenty thousand of them—one person for every hundred yards along the Overland Trail, according to some estimates—very rarely at the hands of Indians, but mostly through accidental gunshot wounds, snakebite, and illness, above all cholera.

Most of them made it. Some five hundred thousand pioneers set out in wagon trains along the Overland Trail from 1841 to 1869, with many splitting off to take the Oregon Trail to the Northwest, and the rest continuing through to California's Sacramento Valley. A few thousand more took the Santa Fe Trail to the Southwest, and seventy thousand Mormons took the Mormon Pioneer Trail to Utah by 1869.

The day of the wagon trains was not long, less than thirty years before they were replaced by the Transcontinental Railroad (see page 48). Yet before they were done they took half a million Americans into the West, walking with their families beside their prairie schooners.

see page 48

THE GENIUS DETAILS

From 1790 to 1840, some four million American pioneers migrated from the Appalachians to the Mississippi.

Only about twenty-five thousand Americans had taken overland routes to the West before gold was discovered in California in 1848.

Typical food provisions of a prairie schooner included six hundred pounds of flour, four hundred pounds of bacon, two hundred pounds of lard, thirty pounds of pilot bread, twenty-five pounds of sugar, ten pounds of salt, ten pounds of rice, five pounds of coffee, two pounds of tea, two pounds of baking soda, one small keg of vinegar, one bushel of dried fruit, one half bushel of cornmeal, one half bushel of parched, ground corn, and one half bushel of dried beans.

One in every five women on the Overland Trail was in some stage of pregnancy. Most families traveled with small children.

Cholera, the deadliest killer on the Western trails, was most often contracted in Nebraska.

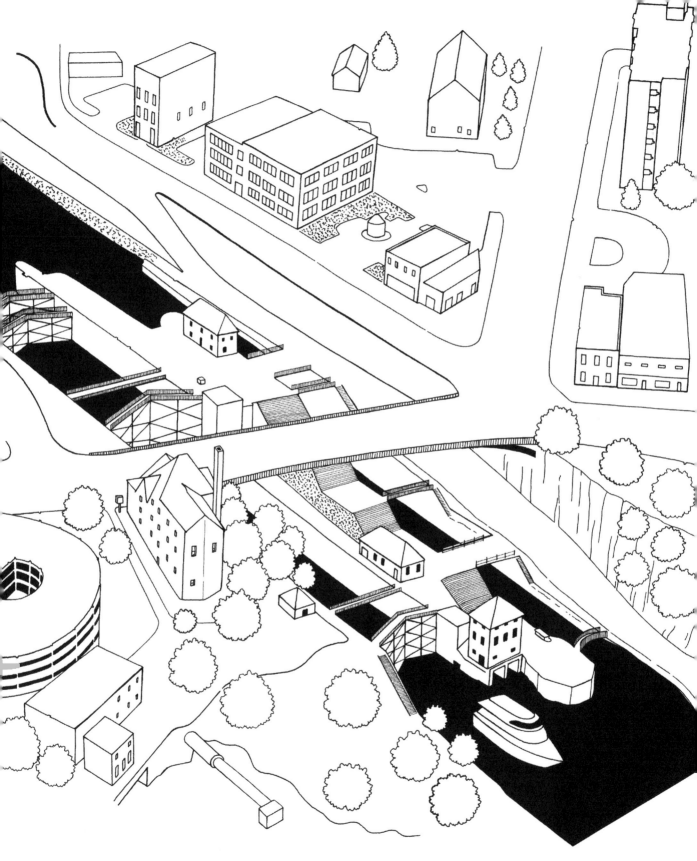

THE ERIE CANAL

N ew York City in the early nineteenth century seemed doubly blessed by geography. It boasted one of the world's greatest natural harbors, and its central location put it closer to the emerging economic colossus of Europe than any other major Atlantic port in America.

The question was how to get the fantastic abundance of raw materials that America had and Europe craved to this port. It was no small sticking point. Roads in upstate New York were little more than old Indian trails, all but impassable much of the year and slowed by outrageous local tolls when they were open. Transporting anything from the rich hinterlands of the continent that America was conquering was infinitely more daunting. Midwestern farmers, miners, and lumbermen looking to bring their goods to market had little choice but to take them down the Mississippi and its tributaries to the Port of New Orleans. Once there, the question was how to get back—especially before the advent of the steamboat. Usually, the answer was to walk (something a young Abe Lincoln did twice), a journey that could take weeks or months.

A solution was proffered in 1807 by one Jesse Hawley, a bankrupt flour merchant from the upstate town of Geneva, with no education beyond a country schoolhouse, languishing in a debtors' prison after failing to get his goods to market before they spoiled. Hawley used his time in stir to write a remarkably sophisticated plan for a "Great Western Canal" system that would connect New York City by water to the shores of Lake Erie.

Similar schemes had been bandied about for decades, but Hawley's had an advantage. It came to the attention of DeWitt Clinton, the volcanic visionary then serving as mayor of New York City.

Clinton had seen New York's potential as a great world city even when it was little more than a pestilential muddle at the toe of Manhattan. Much more, he saw it as a new *kind* of city, one run by free and enlightened men.

To that end, as mayor he fought to abolish slavery and institute universal suffrage, encourage immigration, and end discrimination against Catholics. He pushed through a free public school system, cleaned up New York's filthy streets and markets, and sought to eradicate the diseases that crept up from its docks. He started an orphanage, a literary and philosophical society, and a historical society that flourishes to this day.

The Erie Canal put America—and New York—at the crossroads of the Industrial Revolution when it opened in 1825. Today the canal system's 524 miles carry some commercial traffic but are largely a popular tourist and recreation attraction.

The Erie Canal originally had eighty-three locks to manage a gradual rise of 568 feet in elevation from Albany to Buffalo. The original canal was forty feet wide and four feet deep, with eighteen aqueducts allowing it to cross rivers, streams, and ravines.

The Erie Canal was the greatest public work in the Western world since the Great Pyramid of Giza in 2580 BC. Some 11.4 million cubic yards of earth and rock were removed to build the canal—about three times the amount moved at Giza.

Nearly every major city in New York lies along either the Erie Canal or the Hudson River down to the city. To this day, almost 80 percent of upstate New York's population lives within twenty-five miles of the Erie Canal.

Express passenger service made it possible to get from Buffalo to New York City in an unprecedented four days' time—as opposed to what had been two weeks by wagon or stagecoach just to get from Buffalo to Albany.

From 1824 to 1882, when they were eliminated, tolls on the Erie Canal brought in a total of $121 million, or over seventeen times the canal's original cost.

He also imposed on most of the island the famous "grid system" of straight, numbered streets and avenues that enabled its development.

"Magnus Apollo," as the formidable Clinton was called, was not one to chase moonbeams. But when he saw Hawley's plan he grasped at once that his state contained the only level natural gap through the Appalachian Mountains before Georgia. Cutting a waterway through it to the Great Lakes would connect New York City's harbor to the whole of the Midwest.

Of course, that meant a 363-mile canal system from Lake Erie to the Hudson, where boats could sail the remaining 150 miles down to New York City. The estimated cost was $7 million—or three-fourths of the total federal budget at the time. President Thomas Jefferson thought it might be a good idea a hundred years on but protested, "It is little short of madness to think of it at this day." His successor, James Madison, vetoed any federal funds for it as unconstitutional. Undaunted, Clinton got himself elected governor of New York State in 1817, and within three months of his inauguration shovels were in the ground, the money raised entirely by 6 percent state bonds, guaranteed by the tolls the canal would charge.

Now all Clinton had to do was pull off the greatest public works project in the history of the Western world to date, in a country that barely had a single trained engineer. Most of the Erie Canal was dug out by hand and shovels. Some fifty thousand men worked on it—local farmers, Native Americans, African Americans, German immigrants, and, above all, the Irish. At eighty cents a day, Irish immigrants made five times the wages they could get back home, but contractors fed them swill, housed them in shanties, and dosed them with twelve to twenty ounces of whiskey a day—their only fortification against digging through malarial swamps, quicksand, and icy winter streams, sometimes by torchlight or bonfire. They died in droves from dysentery, yellow fever, typhus, pneumonia, dehydration, falling trees, and faulty equipment and were despised by people in the towns they were about to enrich.

What could not be moved by "Irish power" had to be blown out by volatile black powder in those predynamite days. At the Deep Cut, near Lockport, they had to blast their way through seven miles of rock and wet earth, including three miles of hard blue limestone, to a depth of twenty-six feet. Once the powder was packed into a drill hole, a young boy was used to set the fuse, on the theory that he could run away faster.

By the third summer of work, the area around the Deep Cut resembled a war zone. Workers and hamlet alike were under a continual bombardment of stones, small and huge. An English traveler observed the canal workers to be so fatalistic that

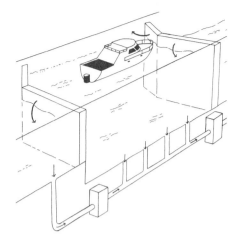

From Albany to Buffalo, the Erie Canal cut a waterway that rose as much as 568 feet above sea level, and fell as much as 363 feet below it. Boats navigated it through locks that were filled or emptied of water to raise or lower the boats as well.

"instead of running to the shelter . . . they would just hold their shovels over their heads to keep off the shower of small stones and be crushed every now and then by a big one."

The first fifteen miles of the canal, from Rome to Utica, was not finished until 1819, and criticism swelled. Clinton's renomination as governor was blocked, and he was removed from his unpaid position on the canal commission. He responded by rallying a broad coalition of voters around him, booted his opponents out of power, and regained the governor's office in time to take a victory lap down the length of the Erie Canal when it opened, culminating in his pouring two barrels from Lake Erie into the Atlantic Ocean on November 4, 1825, in a symbolic "Wedding of the Waters."

Everywhere his constituents greeted him with wild celebrations, as well they might. Not only was the Erie Canal finished at cost and two years ahead of schedule, it would add exponentially to New York's and the country's wealth for generations to come.

Built for an estimated $4.5 billion in current dollars, the canal's debt was retired by its tolls within nine years. Backward villages and towns along its route were turned overnight into bustling industrial cities. The time spent hauling goods from the Midwest to the East Coast dropped from weeks to a matter of days. Shipping costs dropped from $100 a ton to less than $9. New York, now at the cockpit of the industrial world, was well on its way to becoming its leading city—none of which would have surprised the canal's champion.

"The city will, in the course of time, become the granary of the world, the emporium of commerce, the seat of manufactures, the focus of great moneyed operations," DeWitt Clinton predicted. "And before the revolution of a century, the whole island of Manhattan, covered with inhabitants and replenished with a dense population, will constitute one vast city."

THE YANKEE CLIPPERS

They came and went too fast for us to see, save in some old nautical prints and in our dreams. Pillowy forests of sail, moving across the oceans faster than anything ever built. Everything about them was romantic—their names and terminology, the faraway places they traveled to, the elegance of their design. But they were also working ships that sailed through a narrow window in world trade, and when that window closed they vanished. While they lasted, nothing was more beautiful.

Much of America's early mercantile navy consisted of refitted privateers from the Revolution. They were soon joined by "Baltimore clippers," swift little two-masted schooners and brigantines, built to evade British blockaders in the War of 1812 (and to run slaves). They sailed out of New York, Boston, or Salem, Massachusetts, to ports around the world. They were in Canton by 1795, trading silver, ginseng, and furs for Chinese goods that Americans soon could not get enough of: silks and porcelain, cassia and nankeen trousers, lacquerware, fans, furniture, and, above all, *tea*.

The fastest ships had too little cargo space, and bigger ones might take as much as a year out to China and another year back—by which time most teas had lost their flavor. By the 1830s, American shipbuilders up and down the East Coast were working on what we'd call a game changer today: bigger yet still swifter ships that became known as true clippers and then "extreme clippers."

There was no exact definition for what made a clipper. The author Alan Villiers wrote that one "must be sharp-lined, built for speed. She must be tall-sparred and carry the utmost spread of canvas. And she must *use* that sail, day and night, fair weather and foul."

So clippers did. They were long, narrow ships, with deep bows that knifed through the water, and their widest beam moved over halfway back. Their main masts often approached 100 feet in height and ran as high as 230 feet (or twenty-three stories), leaving awestruck observers to dub them "skyscrapers." They were rigged for power, with extra sails everywhere: skysails and moonrakers on the masts, and studding sails on booms, extended from the hull. They ran through everything. In storms where lesser vessels might shorten their sails and try to ride it out, the clippers plunged on into the waves, heeling so steeply their gunwales were in the water.

The *Flying Cloud*, one of Donald McKay's "extreme clippers," that Eleanor Creesy navigated around Cape Horn in record time.

Their speed was unprecedented, covering as much as 465 nautical miles in a twenty-four-hour period, a pace of over nineteen knots. They raced each other around the world, and their arrival was a major spectator sport, with men and women rushing to the waterside to watch them come in. People collected beautifully printed "clipper cards" much as they do baseball cards today.

In the 1840s, the clippers came into great demand as the China trade expanded in the wake of the First Opium War and then as the gold rush brought the world to

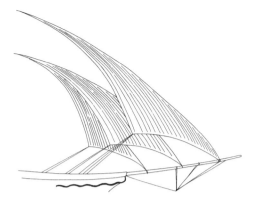

The jib sails at the bowsprit of a Yankee clipper. Clippers boasted extra sails wherever possible, and flew with them hoisted even through the worst weather.

California. Clipper ships were able to halve the six months it had previously taken to sail from New York to San Francisco, and then raced on across the Pacific.

Nearly all the great ship's architects in the United States tried their hand at designing clippers. The best was Donald McKay, a Canadian immigrant who set up shop first in Newburyport, Massachusetts, then down in East Boston. Of the four hundred or so ships that might be called clippers, McKay designed thirty-eight. They had names such as *Staghound*, *Glory of the Seas*, *Westward Ho!*, *Star of Empire*, *Chariot of Fame*, *Lightning*, and *Zephyr*, all monikers that evoked speed or spoke to the spirit of the bold new republic whose wealth he helped amass.

The swiftest of all McKay's creations was *Flying Cloud*, sold for more than $90,000, maybe $20 million in today's money. Its maiden voyage was out of New York, under Captain Josiah Perkins Creesy Jr., from Marblehead, Massachusetts, a renowned maritime town whose sailors had snatched Washington's army from disaster in Brooklyn and rowed it across the Delaware. On board, too, was the captain's wife, thirty-six-year-old Eleanor "Ellen" Prentiss Creesy, which wasn't unusual at the time. What *was* unusual—even unheard of—was that Ellen was also the ship's navigator.

The daughter of a master mariner, Ellen had taken to navigation because she loved the complex mathematics. She learned to use a sextant and a chronometer and studied meteorology, ocean currents, and astronomy. She also pored through the writings of Matthew Fontaine Maury, the dry-docked "Pathfinder of the Seas," whose revolutionary work on sea winds and currents made possible the laying of the first transatlantic cable (see page 71).

The *Flying Cloud* left New York Harbor on June 2, 1851, carrying passengers and 2,000 to 2,500 tons of cargo, including mining supplies and equipment, household goods, cotton duck, and gourmet delicacies for the epicurean prospectors of San Francisco. In twenty-four hours the ship had gone an amazing 389 miles. It crossed the equator in a record seventeen days, at least twenty-one sails hoisted at all times, more when the winds were booming.

All was *not* smooth sailing. Like all clipper ships, the *Flying Cloud* sailed right through storms, In a severe storm off Brazil, it nearly lost its masts. This so frightened some of the crew that they drilled holes in the hull, hoping to force the ship into port. Their sabotage was discovered and repaired—and the *Flying Cloud* flew on. Rounding Cape Horn in winter, it ran into another furious storm. With the sky gone, Ellen Creesy charted their path below deck for one thousand miles by dead reckoning. Her husband stood on the deck for long hours in the wind and freezing rain, the two of them shouting back and forth above the tempest. It was their first time through the treacherous waters of the Horn—but they emerged just where Ellen had calculated, eight miles from land.

The *Flying Cloud* "did the 50-50"—going from fifty degrees south latitude in the South Atlantic to fifty degrees south latitude in the South Pacific—in just seven days, a record that still stands. Relying on Maury's research, Ellen stood the ship farther out to sea off Central America than was customary, and the *Flying Cloud* made it past the Golden Gate in a record eighty-nine days and twenty-one hours, despite losing a mast in yet another storm. In addition to her navigational work, Ellen Creesy had performed the nursing duties expected of a captain's wife and had baked a wedding cake for two passengers who got married on board.

By the Civil War, the era of the great clipper ships was over. Refiguring the eternal calculation of speed versus capacity in maritime trade, shipping lines had begun to switch to somewhat slower but bulkier "medium clippers."

The Creesys had already retired to Salem, where Josiah went into local politics and died in 1871. Ellen lived the rest of her long life far from the sea, though on her death in 1900 her body was brought back to lie beside her husband's in a hillside cemetery in Salem, above the ocean where once they had swept over the waves like a dream.

THE GENIUS DETAILS

The definition of a clipper ship, and particularly an "extreme clipper," has often varied. The first true clipper ship is usually considered to be the *Ann McKim*, nearly five hundred tons, built in Baltimore in 1833.

The size of clipper ships ranged from less than five hundred tons to a maximum of four thousand tons.

It took steamships some twenty-five years before they could surpass the speed of the clipper ship.

Clipper captains were encouraged by their ships' owners to set speed records if they could. As in many maritime endeavors, such as whaling (see page 159), ships' officers were often rewarded with some share in the whole undertaking. Both Capt. Josiah Creesy and navigator Ellen Creesy had a $\frac{1}{32}$ share in the *Flying Cloud*'s initial voyage.

Donald McKay's *Glory of the Seas*, out of his East Boston shipyard in 1869, was the last clipper ship built in America.

THE PANAMA CANAL

A merica had barely made one great surge across the continent before it surpassed itself with another. The Erie Canal cut the time it took to travel from the East Coast to the Great Lakes from two weeks to about four days (see page 9). Within another generation, the Transcontinental Railroad was zipping passengers across North America by train in just three and a half days (see page 48). The brilliant Yankee clippers cut the trip around the Horn from six months to three (see page 13), and in another fifty years, 7,800 miles would be lopped off that voyage. It would take only twenty to thirty hours to pass from the Caribbean to the Pacific—thanks to the Panama Canal.

It was the dream of centuries, to cut a notch through the thin tendril of land that was the Isthmus of Panama. The king of Spain had first ordered it surveyed in 1534, and numerous plans had been proposed over the years. By 1855, a railroad was whisking travelers the forty-eight miles over the isthmus—an engineering marvel that required three hundred bridges and culverts and cost the lives of five thousand to ten thousand laborers. But the journey meant taking your life in your hands, thanks to the "yellow jack," the epidemics of lethal yellow fever that swept the area.

Building a canal proved more difficult than the map suggested. Ferdinand de Lesseps, the French diplomat who had organized the excavation of the Suez Canal, tried to duplicate his feat in Panama in 1881. It was a disaster. By the time it ground to a halt in bankruptcy eight years later, Lesseps's company had run through twenty-two thousand lives, $287 million, and the savings of eight hundred thousand small investors, and Lesseps and his son were both facing five-year prison sentences for immense financial scandals. Their men, most of them poor black workers from the Caribbean islands, had died in droves, not only from yellow fever but from malaria, dysentery, poisonous snakes, jungle insects and spiders, and landslides caused by a rainy season that lasted from May to November. Most of the French engineers recruited for the project had fled back to Europe as soon as they could.

Not Philippe Bunau-Varilla, who had come over to work the canal when he was only twenty-six and who then spent more than a decade lobbying governments around the world to take up the great work again. His plan gained a hearing in 1901, when the youngest and most vigorous president ever to occupy the White House moved in. Teddy Roosevelt believed that cutting a path between the seas was vital to consolidating American economic and naval power. With the help of a New York corporation lawyer named William Nelson Cromwell, Roosevelt got Congress to set up an Isthmian Canal Commission that bought out the remaining equipment and facilities of de Lesseps's company for $40 million, less than half what it was asking.

Secretary of State John Hay inked a treaty with Colombia, which then controlled Panama, to lease the isthmus in perpetuity to the United States in exchange for $10 million up front and then annual payments of $250,000. When the Colombian senate demanded another $10 million at the last minute, Roosevelt sent warships to support a bloodless insurrection by the locals. In the course of four days, November 3 through 6, 1903, Panama declared itself an independent country, appointed Bunau-Varilla its ambassador to the United States, and cut what was essentially the same deal that had been offered to Colombia.

It was an ugly start to a glorious adventure, an imperialist land grab that would leave a lingering resentment in Latin America for decades. The *New York Times* called it "an act of sordid conquest" and the *New York Evening Post* a "vulgar and mercenary venture" (though it was not quite so ugly as it looked: Panamanians had already been actively rebelling against what they considered a distant and indifferent Colombian government for over fifty years).

Digging the canal remained a colossal undertaking that would require over ten years. Just replacing or refurbishing the rotting, tangled equipment and 2,148 buildings the French left behind took months. The Americans brought in state-of-the-art cranes, rock-drilling equipment, and dredges, along with 102 gargantuan, rail-mounted steam shovels—but most of the canal would still be dug by hand, by seventy-five thousand workers in all, drawn from the United States and Panama but also Barbados, China, and Europe, threatened constantly by death from one disease or another.

Yet because of two main innovations, one mechanical and one hygienic, the American effort would succeed where the French had failed.

Unlike the Suez Canal, which was little more than a ditch cut through a desert, the Panama Canal meant digging through nearly impenetrable jungle, up to eighty-five feet above sea level. The Chagres River, where the canal began, could rise by up to thirty-five feet in the seemingly interminable rainy season. US engineers overcame these obstacles by committing to a system of three gigantic sets of locks that would lift ships up and down the hills of the isthmus. This in turn meant creating both the world's largest dam and its largest man-made lake up to that time in order to fill and empty the locks, and it required excavating almost 239 million cubic yards of earth and rock, over and above the thirty million cubic yards already dug out by the French.

If seizing the isthmus in the first place was one of the uglier episodes in the history of US relations with Latin America, the hygienic innovation was one of the noblest. American army doctors under Major Walter Reed had just determined in Cuba, mostly by

Connecting the seas: the Panama Canal today, 48 miles long and serving over 144 world trade routes.

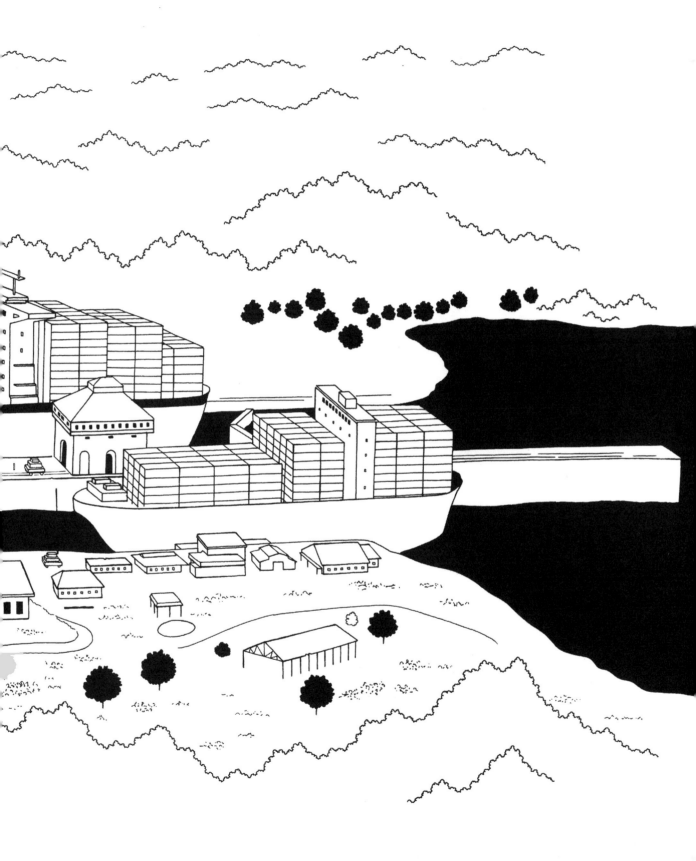

Before efforts were taken to eradicate disease in Panama, 5,600 workers under the US effort would die, from smallpox, typhoid, dysentery, hookworm, and even bubonic plague. In 1906 alone, 80 percent of the workforce was treated for malaria.

For many years, skilled, white American and European workers were placed on the "gold roll"—paid in gold dollars and given first-class food and lodgings to get them to stay. Unskilled workers, especially those of color, were placed on the "silver roll," paid as little as ten cents a day in various currencies, and badly fed, housed, and treated.

The most difficult part of digging the canal was the "Culebra Cut," which went 7.8 miles through a mountain ridge and across the Continental Divide.

The volume of earth excavated for the Panama Canal was over twenty-five times the volume dug out for the Chunnel.

Annual traffic through the Panama Canal went from 1,000 ships in 1914, the year it opened, to 14,702 ships, with 309.6 million tons of cargo, by 2008.

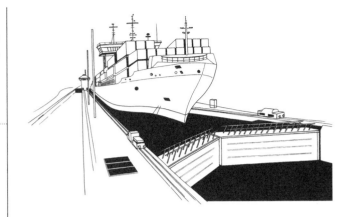

The locks fill with water, then empty again as each ship moves on, lifting the ships up to eighty-five feet above sea level as they cross the Isthmus of Panama. The average toll per ship is $54,000. The largest toll ever paid was $375,000, by a cruise ship, while the lowest was 36 cents, paid by American Richard Halliburton, who swam the length of the canal in 1928.

conducting debilitating and even lethal experiments on themselves, that yellow fever was spread by mosquitoes.

Acting on Reed's results, Major William Crawford Gorgas swept Cuba of mosquitoes wherever possible, all but wiping out yellow fever in Havana. When he arrived to do the same in Panama, Gorgas was mocked by some members of the canal commission. But following a 1914 appeal to Roosevelt, who wasn't in office anymore but was still the most popular man in America, Gorgas's anti-mosquito campaign reduced yellow fever in Panama to half of what it was in the United States itself. And on August 15 of that same year, the SS *Ancon*, a cargo ship, became the first vessel to officially pass through the Panama Canal.

In 1921 the United States agreed to pay Colombia $25 million for the canal rights, and in return Colombia recognized the independence of Panama. Panama has controlled the canal since the end of 1999 and draws $1.4 billion in annual income from it.

BUILDING THE NEW YORK SUBWAY

N ew Yorkers were mad with excitement. All day long, they blasted horns and sirens, rang church bells, fired off guns and cannons, and festooned buildings with flags and bunting. Then, at seven o'clock on the evening of October 27, 1904, over 110,000 of them swarmed underground.

The subway had arrived.

For the rest of the night, they rode the trains for free, many of them lustily singing songs written for the occasion. They rushed off to "subway parties" and danced the "Subway Express Two-Step." The following Sunday, over one million people—nearly three times the system's capacity—tried to "do the subway." They marveled at the speeds at which the trains whisked them through the city and were so outraged when the privately owned Interborough Rapid Transit Company (IRT) started hammering up

A 1 train at today's Times Square-42nd Street station in Manhattan.

A direct-current electric motor, which drew steady electric power through the "third rail," then pumped its energy back into the system every time the trains braked—thereby making possible New York's revolutionary subway system.

tin advertising signs in the stations that the city sued to have them removed. (They were not.)

Americans did not invent rapid, underground transit, but an American *did* provide its most crucial element, and it was in America that the subway first became a viable means of transporting a city's worth of people every day.

By the mid-nineteenth century, the size of cities seemed permanently constricted by the awful knots of traffic on their streets. London was the first to go underground with its "Tuppenny Tube" in 1863. This was a brilliant feat of engineering, but the trains were still pulled by lumbering, coal-powered steam locomotives that spewed smoke, soot, and sparks into the deep tunnels, until one journalist reported he was coughing "like a boy with his first cigar."

The tubes also caught the interest of Frank Julian Sprague of Milford, Connecticut. After losing his mother at eight, Sprague was raised by an aunt in New York City and went to take what he thought was an exam for the US Military Academy at seventeen. It turned out to be a test for the Naval Academy, but he passed with flying colors, and his brief career at sea enabled him to visit the London Underground in 1882. Sprague had already patented his first "dynamoelectric machine" when he was twenty-four, and he quickly grasped how an electric motor might be the solution to running underground trains.

Electric motors had been powering machinery for half a century, but most had proved to be too costly, erratic, and dangerous. Starting his own company with what little financing he could find, Sprague invented the first practical direct-current electric motor. Sprague motors didn't spark, they could be adapted for any size job, and they provided constant, steady power. Soon they were powering factory machines, elevators, and printing presses all over America and Europe.

They were ideal for underground trains. A motor on each car would eliminate smoke and cinders and the locomotive itself. Trains would stop and start much more easily and run more smoothly. Sprague and collaborator William J. Wilgus, who would plan the development of Grand Central Terminal (see page 57), invented a system that allowed his motors to keep touching a "third rail" that conveyed power throughout the system. Whenever the cars braked, their motors *created* electricity that ran back along the extra rail, thus saving nearly three-quarters of the system's energy costs.

Toiling ceaselessly under tremendous time and money pressures to match his motors to an entire transit system, Sprague first electrified the streetcars of hilly Richmond, Virginia, in 1888. London's trains were electrified soon after, and then America's first subway system opened in Boston in 1897.

It was in New York, though, that the subway would reach its apotheosis, becoming almost synonymous with the city. It was built through all manner of terrain: burrowing under riverbeds, soaring over the Manhattan Valley on elevated tracks. The IRT recruited veteran miners from all over the world to dig a two-mile tunnel 180 feet below the surface of Fort George Hill, through Manhattan's treacherous layers of granite schist. In one terrible moment, a three-hundred-ton boulder suddenly dislodged and killed ten men—but the tunnel was built.

Most of the construction utilized a safer technique known as "cut-and-cover," first utilized in Paris. Workers drilled and blasted open city streets, all the while maneuvering carefully among water pipes and power lines and carefully propping up homes, businesses, and even a statue of Christopher Columbus atop a seventy-foot granite column as they worked. Their method was simple. After digging down about twenty feet, they inserted steel columns and beams, then poured in cement stations and tunnels.

The stations became little jewel boxes of craftsmanship, as once again Americans appropriated the best of the Old World, copying the ornate iron-and-glass, Art Nouveau station entrances of the Paris Métro and Budapest Underground. Down below were ticket booths made of polished oak with bronze fittings; leaded-glass skylights and chandeliers; and tiled walls, with distinctive bas-relief panels for each station and mosaic work spelling out each stop's name and street number. All for five cents.

The subway would suffer from neglect and crime periodically over the years, but today it shines brightly, more extensive and integral to the success of the city it serves than ever before.

"Without the subway, New York might very well have turned out to be Bridgeport," wrote urban historian Kenneth T. Jackson.

THE GENIUS DETAILS

Annual ridership by 1930 was 2,049,000,000. In 2014 it was 1,751,287,621.

The subway had 28 stations when it first opened. Today it has 468.

Operating track when the subway first opened extended 9.1 miles. Today it extends 232 miles.

Top speed of express trains in 1904 was forty miles an hour—three times the speed of New York's elevated trains and six times the speed of its streetcars. Top speed today is fifty-five miles an hour.

The three New York subway lines—the IRT (Interborough Rapid Transit Company), BMT (Brooklyn-Manhattan Transit Corporation), and the IND (Independent Subway System)—were unified under public ownership in 1939 in what was the largest railroad merger in US history.

DREAM CAR
THE LINCOLN ZEPHYR

They said he got the idea from watching a flight of geese as they moved through the fall sky in their "V" formation—or a squadron of army planes, or one of the new airships, the Zeppelins, that appeared in American skies in the 1920s. Whichever it was, Carl Breer was convinced that American auto design had to change, and change radically.

The son of a German blacksmith who immigrated to Los Angeles, Breer was inventing from the moment he could walk. Along with his brothers and sisters, he built tricycles, bicycles, a camera, and a wagon their dog could pull. A trip to the Los Angeles Water Works—about to change the whole future of his hometown (see page 215)—left him with a lifelong love of engineering and the friendship of Fred J. Fisher, chief engineer there, who would help Carl build a generator that lit up his whole house. Breer was still in grammar school.

By the time he was eighteen, in 1901, Breer had built his very own steam-powered car from scratch. After college he went to work for a string of car and auto parts companies in the Midwest and on the Pacific Coast, becoming best friends and working partners with two other hugely talented designers and engineers, Fred Zeder and Owen Skelton. "The Three Musketeers" stuck with their boss, veteran auto executive Walter P. Chrysler, when he went out on his own. Together they turned out the first generation of Chryslers—smooth-running, well-engineered beauties that immediately elbowed their way into a significant share of a 1920s auto market that still included forty-four different manufacturers.

Inspired by the skies, Carl Breer wanted something more. Along with Zeder and Skelton, he recruited Orville Wright to help him build a primitive wind tunnel, where they tested at least fifty scale models in April 1930. What they found was startling: American cars, usually built in a "two-box" design, were so aerodynamically retrograde that they were more efficient driven *backward*. When they were full of passengers, 75 percent of their weight rested over their back wheels, wrecking their springs and making them much more dangerous to handle on wet or slippery roads.

"Just think how dumb we have been. All those cars have been running in the wrong direction," Breer remarked.

The most beautiful car in the world? The 1939 Lincoln Zephyr.

The interiors of the first 1934 Airflows featured chrome tubing for the seats and marbled rubber for the floor mats.

Only three 1934 Imperial Airflows are believed to be still extant.

The two great hits of the New York Auto Show in November 1935 were Ford's Lincoln Zephyr and Auburn's Cord 810. Hundreds of spectators crowded around to see them, standing on the bumpers of other vehicles.

Breer would remain at Chrysler until his retirement in 1949. He died in 1970, aged eighty-seven. Edsel Ford would die of stomach cancer in 1943, aged just forty-nine.

In trying to rebut the GM-led attack on its performance, Chrysler put out a short film for theatrical distribution in which an Airflow was shoved off a 110-foot cliff, was righted, then started immediately and was driven away.

The Three Musketeers turned the car around. Utilizing—and inspiring—the same "streamlining" methods then transforming trains (see page 61), Chrysler moved the engine forward and down between the front wheels instead of leaving it in its customary position behind the front axle; evened out weight distribution; made the front seat wider and the rear seat deeper; expanded the front fenders; and built the windshields out of not one but two pieces of glass to form a "V" shape—all to deflect the stream of air *closer* to the car between its wider front and its narrower rear fenders.

In 1934, the Chrysler/DeSoto Airflows were a revolution in design, sleeker, lower, and closer to the ground than anything then on the road, with a full steel body. They nearly wrecked the company, thanks in part to a number of glitches that had escaped notice, such as engines occasionally breaking loose from their mountings when the car reached eighty miles an hour.

Yet the real problem with the Airflow wasn't its performance but its appearance. It may have been the most aerodynamically advanced car in existence, but it didn't *look* fast, thanks to its "waterfall" metal grille, curved hood, and combined form, which led critics to call it an "anonymous lump," or "a lumbering, stupid, almost featureless animal, a blank face with nose and eyes reduced to flat surfaces." Buyers needed two or three days just to "become accustomed to them."

By 1937, the Airflows had been discontinued—but they did not die in vain. They would have a tremendous influence on similar contemporary advances in car design, such as Gordon Buehrig's luscious Cord 810 from Auburn Automobile and the Chevrolet Master Deluxe Sport Coupe.

The car that really fulfilled all the promise of the Airflow, though, was the Ford Lincoln Zephyr. Edsel Ford would endure almost unbelievable cruelty from his father as Henry Ford's mind crumbled—part of what the aging magnate thought of as "toughening up" his son. This would include Henry destroying a car model and coke ovens that Edsel had commissioned, firing a whole department of accountants just to spite him, and hiring a street thug to usurp Edsel's role in most Ford operations (see page 147).

Through it all, Edsel hung on grimly, doing what he could to keep Ford Motor Company, once the wonder of the world, from losing even greater chunks of the car market. He found sanctuary in his design office at the Briggs Body Plant, a major Ford supplier, where he worked for years on a new type of car, one that might outdo all the other astonishing models being turned out by American car companies in the 1930s.

"Father made the most popular car in the world," Edsel Ford once said, "and I would like to make the best."

To that end, he hired John Tjaarda, a Dutch immigrant who had worked extensively on monocoque ("single shell") airplane designs and had built the rear-engine, aerodynamic Briggs Dream Car for the 1933 Ford Pavilion at the Chicago "Century of Progress" World's Fair. Working with Edsel Ford, Tjaarda would produce a series of superb, front-engined variations on the Dream Car, starting with the relatively small 1936 Lincoln Zephyr, which was a huge hit at the New York Auto Show and was immediately declared "the first successfully designed streamlined car in America." A series of Zephyrs—the auspicious name also borne by the most beautiful streamlined trains of the age—followed until Edsel had the 1939 Lincoln Zephyr convertible, a luxury car, built first as a prototype for himself alone. That February he drove it down to Florida, drawing admiring gazes and excited queries from everyone he encountered.

With its "teardrop design" fenders, its long, low lines and jutting hood, its rear trunk scalloped around its spare tire, its gorgeous detailing, and its V-12 "glider-ride" engine, the Zephyr was perhaps the finest combination of style and performance in an American automobile, "a classic car in the age of the classic car," as author Robert Lacey would call it. Henry Ford disparaged it, of course, but the Lincoln Continental would become Edsel's lasting contribution to his company, Ford's flagship car through 2002.

American cars had become something transcendent, as *Arts and Decoration* magazine recognized at the time: "The modern automobile is painting and sculpture in motion."

THE TRANSCONTINENTAL PLANE

T he tragedy made headlines all over the United States: KNUTE ROCKNE KILLED AS AIR LINER CRASHES. It was, President Herbert Hoover wrote in a telegram to his widow, "a national loss." Coach Rockne was already a legend, the Notre Dame football coach who would always be remembered for his "Win one for the Gipper" halftime speech. His death in a Kansas field, on the last day of March 1931—along with all seven of his fellow passengers and crew members—sent shock waves around the country.

At the time, plane travel had taken tremendous leaps and bounds, but it was still a mode of transportation primarily for those who liked to take their lives in their hands. The probe into Rockne's death revealed a big reason why. The plane he was traveling in, the Fokker Super Universal Tri-Motor, was considered a piece of cutting-edge technology, fast, light, efficient, and capable of going faster than 120 miles an hour with a 700-mile range. But the investigation by the Aeronautics Branch of the federal Department of Commerce revealed that the wooden structure that made the Fokkers so light also made them death traps. The plane Coach Rockne had flown in was rotted through, the panels of one wing separated by moisture that had seeped inside it undetected.

The wooden Fokkers were immediately grounded. The fledgling Kansas City–based airline Rockne had been flying, an outfit known as Transcontinental & Western Air, or TWA, tried to replace them with the metal Ford Tri-Motor, but this was a slow plane that looked all too much like the Fokker. Desperate, TWA turned to the Boeing Company, which had just launched its new 247 line, the first real modern airliner, with an all-metal, stressed skin, radial engines, and retractable landing gear. But Boeing's parent company, United Aircraft and Transport Corporation, also owned United Air Lines and had promised to give the 247 exclusively to United for a year.

TWA turned to one contractor after another, looking for someone to build them a plane. The man they found was Donald Douglas, who had already decorated his Douglas Aircraft plant in Santa Monica with a giant picture of the Boeing 247 and the caption "Don't copy it, do it better!"

Born in Brooklyn, Donald had followed an older brother into the US Naval Academy, but he was never able to overcome his lifelong obsession with planes. Even

The DC-3 sparkles in the air in 1989, over fifty years after its introduction—its burnished silver skin, in the words of architectural historian Richard Guy Wilson, "creating a new standard of machine beauty."

Eddie Rickenbacker, America's leading flying ace in World War I, headed a team that flew the DC-1 from Burbank, California, to Newark, New Jersey, in a then record thirteen hours and four minutes, to create publicity for TWA.

The DC-3 first used a Wright R-1820 Cyclone 9 engine before shifting to a Pratt & Whitney R-1830 Twin Wasp for better altitude and performance.

Douglas Aircraft was acquired by Howard Hughes in 1941. It would remain the nation's leading aircraft maker during the propeller era but would be surpassed by Boeing in the jet age.

The Douglas-Boeing competition would spur great advances in aviation until Douglas Aircraft was acquired by McDonnell Aircraft in 1967. McDonnell Douglas would merge with Boeing in 1997.

The death of Knute Rockne in a 1931 airplane crash would lead to a federal takeover of air safety standards and inspections.

at Annapolis, he built model aircraft in his dorm room. Dropping out, he enrolled at MIT and became the first person to graduate from the school with a degree in aeronautical engineering, completing his undergraduate work in just two years. He quickly found positions at different firms in the emerging industry, then started his own company, trying to produce the first plane that could fly nonstop across the North American continent, the Douglas Cloudster.

The Cloudster was downed by engine trouble, but Douglas went on to found another company, concentrating on selling torpedo planes to the navy, and eventually picked up a company belonging to a talented draftsman named Jack Northrop. Northrop helped Douglas produce six Douglas World Cruisers—biplanes with flotation devices—for the army, planes with the ability to fly around the world. But Douglas wanted more.

What TWA wanted was a three-engine aircraft of all-metal construction that could take off from any field the airline owned and could seat—and sleep—twelve. Douglas flew his whole engineering staff to New York to show the airline he was serious and convince TWA that the three-engine plane was a thing of the past. Instead, using design advances initiated by Northrop, Douglas produced a prototype of a DC-1—Douglas Commercial-1—for 1934 with more powerful and efficient twin propeller engines. From there he moved on quickly to the DC-2, with its all-metal, multicellular wing stuck under the fuselage, where it gave stronger support and allowed more room for the passengers. (The Boeing 247 was so crowded on the inside that passengers had to either climb over or sit astride the huge wing, which ran right through their compartment.)

In 1936, though, Douglas came up with his greatest plane of all: the DC-3. The first airliner ever to make a profit from carrying passengers alone, it "revolutionized air transport in the 1930s and 1940s," according to transportation historians, and was "one of the most significant transport aircraft ever made." Efficient, reliable, and easy to maintain, it could use the shortest TWA runway.

An aviation version of the stream-lined cars (see page 25) and trains (see page 61) of the era, tested in Caltech wind tunnels, it could attain a speed of up to 207 miles per hour and a range of 1,500 miles. Where the Transcontinental Railroad had taken passengers across America in 83 hours back in 1876, the DC-3 could do it in just 15 hours east to west (17.5 hours west to east), with a couple hops to refuel.

This speed and range was crucial. It meant the sleeper berths could be removed, allowing the DC-3 to comfortably—and profitably—carry a record twenty-eight

The DC-3 was so fast that the traditional sleeper berths on long-distance planes weren't necessary and could be removed, providing passengers with unprecedented space and comfort.

passengers inside its stressed-skin, semimonocoque, light aluminum body. But the key selling point was not how it performed but how gorgeous it looked. The DC-3's burnished silver skin was arranged in squares that "flowed into each other, creating a new standard of machine beauty."

The competition was routed. DC-3s were bought by American, United, Piedmont, and Eastern Air Lines, as well as TWA. They were adapted or purchased by countries from Japan to the Soviet Union, China to Cuba, and were pressed into military service as the C-47 Skytrain during World War II. By that time, Donald Douglas was busy making the SBD Dauntless Dive Bombers, which would turn the course of the war in the Pacific in a matter of minutes at Midway. And the DC-3 . . . never really did go away. An estimated four hundred of them are flying regularly scheduled passenger flights to this day, and it may become the first plane ever to last a hundred years in service. It was just done better.

THE CONTAINER SHIP

By 1956, the Port of New York had been the busiest harbor in the world for at least half a century—and it was moving backward. Though it was located beside the greatest city in the most technologically advanced nation on earth, unloading a ton of cargo there took 50 percent *longer* than it had just six years earlier. Pilferage was rife, culminating in the disappearance of an entire electrical generator from the docks. The amazing hedgehog of 283 wharves that rimmed Brooklyn and the lower half of Manhattan Island, some of them as much as seven hundred feet long, was crumbling from long neglect.

The fifty thousand men who worked "alongshore" to load and unload ships did backbreaking work with the most minimal of tools, but they, too, were the victims of abuse, savagely exploited and subjected to the notorious "shake-up" that forced them to practically beg for work every morning. They were slaves to a corrupted union controlled by a single, shady businessman, one William McCormack, who employed some of the most notorious Mob killers in American history to enforce discipline on the docks and beat or murder anyone who dared to stand up to him. Out of fear of these goons or a misguided sense of solidarity, almost no longshoreman would testify against such tactics. The most famous Mob-busting prosecutors in the country, fearless political reformers, a Pulitzer Prize–winning exposé in the *New York Sun*, a contingent of battling labor priests, and acclaimed books, plays, and movies by some of America's leading artists—nothing was able to break the hold that one evil man had on the nation's greatest port.

Yet McCormack, the Mob, and their ilk proved no match for another solitary, determined individual from Maxton, North Carolina, Malcom McLean.

In the postwar years, it took an average of eight days to load "break-bulk shipping"—anything in a crate, a box, or a bale—onto a ship, and eight days to unload it. This was largely true even on the West Coast, where a breakaway waterfront union under radical Harry Bridges had greatly diminished theft and improved working conditions. Going much faster simply wasn't possible; one small cargo ship of 5,015 tons, for instance, was loaded with 194,582 separate items.

One of an estimated 5,000 container ships at work today, hauling some 300 million shipping containers in use around the world.

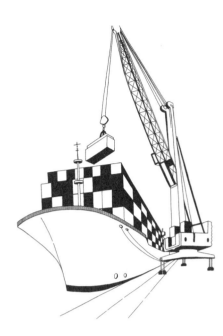

A gantry crane loading directly onto a container ship. Worldwide, container ships cut the cost of shipping an object from 50 percent to just 10 percent of its price.

McLean had a better idea, startling in its simplicity: put the truck trailer on the ship. Standardized containers had already been tried with considerable success during World War II. Andrew Jackson Higgins, the visionary New Orleans boatbuilder who manufactured over twenty thousand landing craft for the United States with an integrated workforce, and "who won the war for us," according to Eisenhower, had tried to emulate the idea in civilian life, but he had found no backers and his health broke down.

McLean, on the other hand, was in more of a hurry than ever. A chain-smoking executive with a head for numbers and a penchant for taking risks, McLean had begun his working life during the Great Depression, hauling dirt and tobacco around his hometown of Maxton, North Carolina, with a $120 used pickup. He soon became a long-haul trucker of cotton up to the New York docks.

By 1953, when he was still just forty, McLean had—with his sister Clara and his brother Jim—built his trucking business into a $12-million concern, operating 1,700 trucks, but he threw it all over, convincing National City Bank (later Citibank) to help him buy a small shipbuilding company and a couple of used, oceangoing oil tankers. His first ship, the *Ideal-X*, was reconfigured by engineer Keith Tantlinger, who designed a new aluminum container and invented the "container spreader bar," which let the containers be lifted and lowered by cranes without any ropes needing to be attached. This meant, as the historian and economist Marc Levinson has written, that "once the box had been lifted and moved, another flip of the switch would disengage the hooks, without a worker on the ground touching the container."

The standardized aluminum or steel containers could be stored anywhere on a ship, even the open top deck. There was little chance for thieves to break in or even know what was inside. At the end of the voyage, each container could again be lifted and lowered directly onto a freight train or a truck chassis. A high-speed crane would soon be able to load a twenty-ton container every three minutes, making it over forty times more productive than a longshoremen's crew with their winches, pallets, and hooks.

The SS *Ideal-X*, nicknamed the SS *Maxton*, set sail from the Port of Newark for Houston with fifty-eight containers and its regular load of liquid tank cargo on April 26, 1956.

"Containerization" would fundamentally alter both the way world trade was conducted and what American cities would be, for better and for worse. McLean's bold innovation made it possible to move goods from developing countries around the world to the United States and Europe. But as the geographer David Harvey has pointed out, they would also spur the massive "deindustrialization" of America.

Container ports, requiring vastly more open land and water to operate, were now built outside major metropolises, all but annihilating cities' advantages as manufacturing centers. By 1958, for example, just two years after Malcom McLean's first ship set sail, the Port Authority of New York and New Jersey started building the massive Port Newark–Elizabeth Marine Terminal out by the Jersey Meadowlands. It would crush the power of the port's corrupted unions and—a generation later—would lead to a massive real estate boom as New Yorkers found their way back to the waterfront. Container ships would also provide an increasingly energy-efficient way to move goods around a warming planet. In several parts of the world, sunken containers have been used to construct artificial ocean reefs. But at the same time, new container ports around the country eliminated millions of decent working-class manufacturing jobs, traditional stepping-stones into the middle class. The change helped destabilize American cities, then left them seriously divided between the haves and have-nots. Like so many new technologies, the container ship was probably inevitable and usually beneficial. But we have still to adjust to what it wrought.

THE GENIUS DETAILS

The biggest container ships today are usually about 1,300 feet long and 180 feet wide. They can be as much as twenty-one stories tall, with engines that weigh 2,300 tons and propellers that weigh 1,300 tons. They can now carry up to 18,270 TEUs.

A fully loaded coal train—the heaviest and largest form of land transportation today—carries twenty-three thousand tons. A large container ship carries three to four times that weight.

Modern container ships have an average life span of twenty-six years. Their fuel efficiency improved by 35 percent between 1970 and 2008, and they emit on average roughly forty times less carbon dioxide than a large freight aircraft and three times less than a heavy truck. Overall, container shipping is 2.5 times more energy efficient than rail shipping and seven times more efficient than shipping by truck.

The computer tracking of container ships is now so precise that a two-week voyage can be estimated for arrival within a range of fifteen minutes. Nearly all containers are bar-coded.

APOLLO 11 AND THE SPACE RACE

If they could put a man on the moon, why can't . . ." began the cliché after July 20, 1969, referring to the most prodigious undertaking anyone could imagine. But to this day most Americans probably do not understand what it took to get a dozen men to the moon and bring them back again.

Putting a man on the moon would take eight years of work, including ten manned and unmanned missions. It would mean building a huge new Mission Control Center for the National Aeronautics and Space Administration (NASA) in Houston and a new launch center at Cape Canaveral. But by 1969, Apollo 11, the remarkable spacecraft that would take humankind to the moon, was ready. (The program was dubbed Apollo by NASA manager Abe Silverstein because "Apollo riding his chariot across the Sun was appropriate to the grand scale of the proposed program.")

It was really three spacecrafts in one, a design from a NASA engineering team headed by Maxime Faget, "a truly essential genius of the early U.S. space program," according to author Andrew Chaikin.

First, there was the Saturn V rocket, developed by former-Nazi rocket scientist Dr. Wernher von Braun. It was the largest and most powerful rocket ever built, a brilliant, complicated piece of technology with 950,000 gallons of fuel to blast the whole ship into orbit, 118 miles above the earth. After a turn and a half around the planet, its rockets were fired again, slingshotting Apollo on its translunar trip.

After that, "all" that the three astronauts aboard—mission commander Neil Armstrong and pilots Edwin "Buzz" Aldrin and Michael Collins—had to do was detach their command module, *Columbia*, from the Saturn V, turn it around in space, attach the lunar module *Eagle* to itself, unlock the *Eagle* from the spent rocket, then speed on to the moon.

After Apollo 11 reached lunar orbit on July 19, Armstrong and Aldrin undocked the *Eagle* from the command module and headed for the moon's surface. But three

The Apollo 11 spaceship sits atop its mighty Saturn V rocket, with its 7.5 million pounds of liftoff thrust.

minutes from landing, the lunar module's computer started to sound an alarm and flash error message numbers unfamiliar to the astronauts—or anyone at Mission Control. Deciphered, the computer's alarm signal was saying it was overloaded. The trouble was a radar switch set in the wrong position.

The *Eagle*'s computer possessed thirty-six kilobytes of memory, or less than your cell phone uses today. It was, however, incredibly reliable, thanks to the crack team of MIT scientists who had literally hardwired it. Margaret Heafield Hamilton, a thirty-three-year-old pioneer in software engineering (she had, in fact, coined the term *software engineering*), had also incorporated a set of recovery programs into the *Eagle*'s software so that during the descent to the moon it could repeatedly reboot itself and reprioritize its tasks.

Seconds later, though, as the *Eagle* zipped along four hundred feet above the moon's surface, a new problem emerged. The planned landing site was a spot on the Sea of Tranquility, a smooth lava plain. Passing before the window of Commander Armstrong was, instead, a seemingly endless field of crater holes and boulders the size of trucks. When the *Eagle* had undocked from the command module, its cabin had not been fully depressurized, releasing a small burst of gas that sent it a critical four miles off course.

Armstrong, a navy combat and test pilot, took manual control of landing the *Eagle* with just sixty seconds of fuel remaining for the descent. Back in Houston, the mission's controllers wondered if they should order the mission aborted—a reversal that, with the *Eagle* so close to the surface, might easily have killed the astronauts. Ten feet from the moon, the ship was kicking up so much dust that Armstrong had to gauge his landing by some nearby boulders and by the shadows cast by the sun. He did so perfectly. The "small step for a man" actually turned out to be a three-and-a-half-foot jump; Armstrong had landed his craft so gently that its shock absorbers had failed to bend.

"Houston, Tranquility Base here. The *Eagle* has landed," Armstrong radioed back calmly, a report heard by an estimated six hundred million people, or one-fifth of the world's population.

Yet even now the astronauts were far from "safe" in such an unforgiving environment. After they landed, the extreme cold from the moon's surface permeated a fuel line, plugging it with ice. Realizing this had the potential to cause a massive explosion, Houston scrambled to decide whether the *Eagle* should immediately lift off again.

Instead, the heat of the engines melted the ice, relieving the pressure. But when the astronauts were ready to walk outside, there was still too much air in the capsule to depressurize it and get the door open. Aldrin had to carefully peel back an edge of the hatch. Then, as Armstrong struggled to squeeze through, his backpack snapped off the arming switch for the ascent engines—a tiny piece of equipment that might have proved irreplaceable, 239,000 miles from a hardware store.

Armstrong would later replace the switch and arm the engines with his Fisher Space Pen. In the meantime, he walked down the ladder, stepped onto the moon, and uttered the historic phrase, "That's one small step for a man, one giant leap for mankind."

Or something like that. An interruption in the transmission may have obliterated the first "a," or he may have forgotten to include it. Incredibly enough, Armstrong's words were not prescripted. He had only bothered to think them up *after* the *Eagle* had landed, believing that "the chances of a successful touchdown on the moon surface were about even money—fifty-fifty. . . . So it didn't seem to me there was much point in thinking of something to say if we'd have to abort landing."

The astronauts' time on the moon's surface was limited because NASA did not really know how well their spacesuits would hold up in the moon's atmosphere. There was even some speculation that once the moon dust on their suits mixed with the oxygen back in their module it would cause a conflagration.

Armstrong and Aldrin spent about two hours on the powdery bed of the Sea of Tranquility, gathering moon rocks and data, before blasting off again to rendezvous with Collins, who had remained on the command module—out of human contact whenever his craft swung around to the dark side of the moon. But a great challenge remained: reentry to the earth's atmosphere, at speeds that would reach seventeen thousand miles an hour and heat their craft to nearly four thousand degrees Fahrenheit.

Fortunately, NASA engineer Harvey Allen had designed the *Columbia* to look "like an inverted Styrofoam coffee cup with a saucer-shaped lid," in Chaikin's description. Entering the atmosphere blunt end first, it created a shock wave that directed most of the heat away from the capsule. The command module splashed down safely in the Pacific, not far from Wake Island. The total duration of the Apollo 11 mission was eight days, three hours, eighteen minutes, and thirty-five seconds.

There would be six more Apollo missions, with five of them landing men on the moon. The project would lead to huge gains in computer technology, but above all it was a story of ingenuity, teamwork, and unsurpassed courage. Forty years after Apollo 11, Neil Armstrong's footprint is still visible on the moon, and it may be there a million years from now—a final testament to what a forbidding and alien environment mankind ventured into.

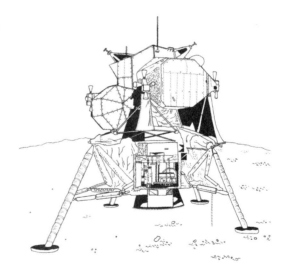

The lunar module *Eagle*, the first spacecraft to land on the moon.

HOW WILL WE TRAVEL IN THE FUTURE?
FROM THE SPACE ELEVATOR TO THE HOVERBOARD

"**I**t's like we're living in the fifties here," Jerry Seinfeld once griped about the lack of flying cars, the advance whose advent was so widely heralded that its failure to materialize has become a running cultural joke. At long last, the flying car may be winging its way to you—but don't expect the model George Jetson used to tool around in.

The first patent for a flying car was registered back in 1903, and aviation pioneer Glenn Curtiss's Autoplane—essentially a Model T with wings and a propeller attached—made its debut in 1917. The years since have witnessed some 2,400 serious flying car designs, with at least 300 of them able to get off the ground. But nearly every model looked pretty much like either a small airplane with wheels—such as Robert Edison Fulton Jr.'s Airphibian—or a car with wings attached—such as the Henry Dreyfuss ConvAirCar. They came with high price tags and a dismaying record of fatal crashes. Even the US military, with its very deep pockets, abandoned attempts in the 1950s to develop a "flying Jeep" for the battlefield.

The intrinsic problem is that an aircraft needs to be as light as possible, while a car needs to be solid enough to withstand gravity-bound wear and tear, not to mention crashes. Attempts to reconcile these disparate needs have generally resulted in building both a bad plane and a bad car.

This may be changing. Taking advantage of technological advances in fuels, engines, and materials, companies in the United States and Europe are trying to develop a new generation of flying cars—or "roadable aircraft"—that look as enticingly cool as anything out of *Back to the Future*, with dual hulls or even flying saucer shapes. Some, such as American Paul Moller's Skycar 200/400, can operate on electricity plus pretty much any other fuel and use only three wheels, making them officially more flying motorcycles than cars. Others are more like helicopters: the Netherlands' PAL-V One is, for example, also on three wheels, with a rotor that has no stall speed, so that even if the engine gives out your vehicle will land softly. While these and other prototypes

A "space elevator" climbs high above the earth.

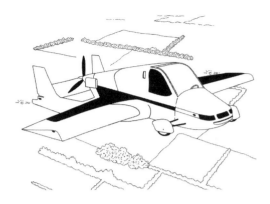

The Transition roadable airplane, with foldable wings and a pusher propeller, currently under development by Terrafugia, an American company founded by a group of MIT graduates.

are currently priced at anywhere from $246,000 to $500,000, mass production could conceivably drive the price down to $50,000, cheaper than some luxury cars.

Don't start looking for a local dealership just yet. Problems big and small remain. Many of the new generation of flying cars still have wings or rotors that need to be manually folded away, making the transition from air to land far from seamless. And don't think you'll just be taking off from your next traffic jam. Most of the prototypes still require at least a short runway, meaning you'll need either a large estate or one of our already overcrowded airports. Even those with a vertical lift can hardly be put down in the driveway; it would be like "sandblasting the neighbor's car while landing on a column of noise on the front lawn," noted John Brown, program manager for Germany's Carplane.

Reducing the sound and the fury from a takeoff might prove possible, but there remain serious questions about keeping order in the air. Picture a bottomless six-lane highway operating just over your house, and you have some idea of what sort of chaos unregulated carplaning might produce. To prevent sheer carnage in the skies (and on the ground below), flying cars would almost certainly have to be *self*-flying cars, directed by intricate computer systems to keep from constantly colliding and plummeting to earth. What should happen if the computers fail—or are sabotaged—is anybody's guess.

"Mark my words: a combination airplane and motorcar is coming. You may smile, but it will come," Henry Ford said—in 1940. We're still smiling. But we may see flying cars in at least some capacity—as emergency service vehicles, air ambulances, or even vehicles for daily commutes by the very rich—before another lifetime goes by.

Why fly, though, when you can take the elevator? The space elevator, that is.

The idea of shooting a sort of endless rope into outer space was thought up all the way back in 1895, by Polish-Russian scientist Konstantin Tsiolkovsky, and was popularized by Arthur C. Clarke's 1978 science fiction novel *Fountains of Paradise*. Tsiolkovsky imagined connecting a cable from a "celestial castle" to the top of the Eiffel Tower. A group of four American engineers revived the idea in 1966 as a "skyhook" into space, complete with a space elevator to take payloads up and down—a concept refined in 1975 by Jerome Pearson, a Marine Corps veteran, engineer, geologist, and inventor working for the Air Force Research Laboratory.

Pearson, looking for a way to slash the cost of getting equipment into space for NASA, and its new space station, conceived of this space "tether" as extending all the way to a counterweight—a man-made space station, or even a captured asteroid—maybe 89,000 miles into space, or nearly half the distance to the moon. This cable would be narrowest on either end, tapering to its thickest at geostationary, or geosynchronous, orbit—that is, 22,236 miles above the earth—where the tensions on it would also be greatest.

At the time, these ideas looked no more realizable than Tsiolkovsky's celestial castle, mostly because no material on earth was suitable for building the skyhook. Steel was not remotely strong enough, and diamonds were too brittle. But the development of nanotechnology in this century may well change everything. Lightweight carbon nanotubes are a hundred times stronger than steel, more than able to stand the stress on such a rope to the stars. Diamond nanotubes look just as promising.

Physicist Bradley C. Edwards has suggested that such materials could be shaped into a paper-thin ribbon 62,000 miles long. Tethered preferably to an island mountain, or some sort of floating station near the equator in the western Pacific—where the risk from hurricanes, tornadoes, and lightning would be smallest—the space elevator cable would rotate with the rotation of the earth. The higher it got, the less the drag of the gravitational force weighing it down, and the higher the centrifugal force pulling it upward, until the two forces balanced at geosynchronous orbit. The tensions on the cable would be greatest there, meaning the "rope" would have to gradually widen until it reached that point, then taper off again as it continued upward to its (asteroid?) anchor.

Unlike building elevators, the space elevator would not have moving cables. Instead, its "cars" would likely be vehicles held to the cable by rollers, or magnetic levitation technology, and would be powered by either laser beams fired at photovoltaic cells or by solar or nuclear energy. It could be built by methods not unlike those used to construct suspension bridges, with the cable literally unspooled—both down to earth and up into space—from a spacecraft at the point of geosynchronous orbit. Every subsequent elevator could be built all the more easily, by materials simply hoisted up the first cable.

How likely are we to see such a lift to the stars? Very, if the nanotechnology can be perfected. The cost of building such a structure is currently estimated at anywhere from $6.2 billion to $20 billion. But it would cut the cost of putting into orbit all sorts of payloads, for commercial or scientific endeavors, from the $11,000 per pound that firing them in rockets now costs to as little as $100 a pound. Even this cost might be quickly defrayed by "space tourism." The speed of climbing the elevator might have to be "limited" to 190 miles an hour in order to keep the cable intact. That would mean a ten-day space cruise into orbit and back as you watched the big, blue marble of the earth open up beneath you.

Plenty of challenges still have to be met. Any "elevator to the stars" would have to withstand threats from weather, man-made space debris, and other hazards, but ascending it would still be exponentially safer—and more comfortable—than blasting off in a rocket.

As in all things, it's likely that developments in the technology of some modes of transportation will spur advances in *all* of them. The magnetic levitation vehicles that will probably climb the space elevator, for instance, are now being perfected considerably closer to earth. They're called "trains."

The use of magnetic levitation, or "maglev," technology—the use of magnetic fields to both suspend objects above the ground and push them forward—was envisioned by a number of scientists in the United States and Europe at the turn into the twentieth century. In 1912, Emile Bachelet, a largely self-taught native of France who had been orphaned at the age of nine and grew up with his younger brother on the streets of Paris before immigrating to the United States, developed the first miniature demonstration model of how such a train might run. When he took his invention to England in an effort to attract investors, Winston Churchill called it the most wonderful thing he had ever seen. But the inability to successfully generate and maintain the level of electrical power necessary for his magnets ultimately doomed Bachelet's venture.

In 1960, the American nuclear physicist Dr. James Powell came up with a new version of a maglev train while—unsurprisingly—stuck in traffic on the Throgs Neck Bridge. Together with his colleague Dr. Gordon Danby, Powell worked out a concept whereby superconducting magnets facing each other on a train and its track would lift the train off the ground and propel it forward as it passed through a series of magnetic loops. Such a train would be freed from the friction of wheels meeting rail. It probably wouldn't even *have* wheels, slowed only by air resistance, as if it were a plane.

Over the next five decades, Dr. Powell and Dr. Danby's ideas transformed train travel as we know it . . . but not in America. Working maglev trains in Japan now routinely reach speeds of up to 311 miles an hour, and have made test runs as high as 375 miles an hour. Maglev trains already reach comparable speeds in China, Taiwan, Korea, and Germany.

Don't plan on stepping aboard one anytime soon in the United States, though, where intercity express trains now dawdle along at an average of 69 miles an hour, slower than they traveled in the 1930s. Maglev trains require dedicated tracks, whereas Amtrak passenger trains must share the rails with freight trains. Reclaiming abandoned rail corridors in heavily trafficked areas such as the Northeast and building maglev tracks is estimated to be prohibitively expensive—perhaps as much as $150 billion—if the right-of-ways can be regained at all.

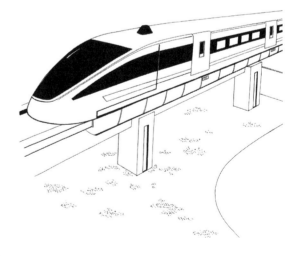

Magnetic levitation trains have attained speeds of up to 375 miles an hour.

Building such lines through the less inhabited West would be considerably cheaper. But America's sheer size would likely make any such project impractical. Cutting the roughly forty-eight hours it currently takes to get from Chicago to the West Coast to, say, eight hours . . . would still take several times longer than a commercial flight would.

Then there's the hoverboard. The idea of hovercrafts—vehicles creating their own layer of air to ride on—has been around since at least the early eighteenth century. Successful oceangoing hovercraft vessels were produced beginning in the 1960s, and some working prototypes of "hovertrains"

Straight from *Back to the Future*, the hoverboard—no, not that silly scooter thing, but a *real* hovering hoverboard—may soon be an ordinary plaything.

or "aerotrains" have also been produced. But hoverboards—skateboards without wheels, that operate a few inches above the ground—first captured America's imagination in, once again, those visionary masterpieces, the *Back to the Future* movies, and their magic has survived segues into Segways and various hoaxes.

Several inventors claim to have invented some version of a hoverboard, including Greg Henderson's Hendo company, which seems to rely on conductors and an oscillating magnetic field. Lexus recently produced a hoverboard it called SLIDE, complete with smokin' cool liquid nitrogen—basically dry ice—wafting from it. SLIDE, introduced in a car ad that had nothing to do with the hoverboard, operates on maglev technology, with superconducting magnets repelling each other to keep the board afloat. But for them to be effective, the surface underneath must contain some magnetic materials. Cue the hoverboard parks, and maybe even hoverboard sidewalk/street strips, but don't count on anything more, anytime soon.

That leaves us with . . . cars. Not flying cars, but good old driving cars. Cars, in fact, that drive themselves.

Self-driving cars are not only the transportation of the future, they're already here, legal for at least road testing in California, Florida, Michigan, Nevada, Idaho, and Washington, D.C. Google has been testing its self-driving "Google Chauffeur" systems for years now, utilizing a Toyota Prius, an Audi, a Lexus and a Lexus SUV, and its own very compact Google X cars, which look like something out of Woody Allen's futuristic comedy *Sleeper*. All told, the cars have undergone over a million miles of road and highway testing and have endured only a handful of minor accidents, nearly all of them caused by people-operated cars rear-ending or side-swiping them. Delphi, the auto parts supplier, has had similar success, sending its self-driving Roadrunner car 3,400 miles across the country, from San Francisco to New York, without incident.

The "Very High Speed Transit System" (VHTS), conceived by the American engineer and nuclear physicist Robert M. Salter Jr., proposed a system whereby people might travel as fast as 14,000 miles an hour. Salter calculated that a capsule holding 100 individuals could go from New York to Los Angeles in just twenty-one minutes.

Daryl Oster's Evacuated Tube Transportation Technology (ET3), or "Space Travel on Earth," envisioned using maglev technology inside near-vacuum tubes to whisk people around the United States at speeds of 370 miles an hour, or at 4,000 miles an hour for international trips. New York to Los Angeles in forty-five minutes? New York to Beijing in two hours?

The American immigrant and inventor Elon Musk proposed the Hyperloop, which he described as a "cross between a Concorde and a railgun and an air hockey table." The Hyperloop would be a pair of elevated tubes whisking passengers from Los Angeles to San Francisco in thirty-five minutes, reaching a maximum speed of 760 miles an hour. Its claimed cost would be no more than $6 billion.

Self-driving cars are equipped with cameras that constantly monitor the car's surrounding environment—more quickly, accurately, and attentively than any human drivers do. Lasers also pick up traffic lights and signs, map the passing scene, and compare it on an inch-by-inch basis with maps from the vehicle's built-in global positioning system. Sensors monitor speed and pick up the nearness of other cars and pedestrians, and a radar system detects many threats before people can possibly become aware of them, such as that bicyclist about to cut out from behind a high hedge, or an accident ahead of that semi blocking your view. A central computer puts it all together and steers, brakes, or accelerates more quickly and adroitly than your basic NASCAR star.

Currently, the cost of all this hardware and software puts the cost of a self-driving car at over $300,000. The sensors on the Google X alone cost $70,000. But these prices are likely to come down radically as computer costs continue to drop and as auto companies around the world perfect and add "autonomous" features, such as advanced cruise control, "automatic parking," and warning systems, to old-fashioned, people-driven cars.

Meanwhile, a world full of self-driving cars would save us many billions of dollars as a society—not to mention untold grief and suffering. According to estimates quoted in the *Washington Post*, a United States in which 90 percent of cars were "autonomous" vehicles would save 21,700 lives every year from auto accidents. There would be over four million fewer accidents and over $447 *billion* in annual, comprehensive savings, figuring in all those smash-ups, deaths, injuries, and other factors. Morgan Stanley estimates that autonomous vehicles would save the United States $170 billion every year in lower fuel costs and another $138 billion in reduced congestion, and having to buy auto insurance might well become a thing of the past. Converting just *10 percent* of American vehicles to self-driving cars would save $37 billion and result in 1,100 fewer road deaths.

What's more, self-driving cars would not require the building of some vast new infra-structure or the development of new technologies. The infrastructure to accommodate such cars exists now, in our roads and highways—though eventually wireless induction chargers might be imbedded there, enabling the coils within your car to recharge your electric motor as you drive.

Unlike maglev or vacuum trains, autonomous vehicles can be added gradually to our everyday lives, an element critical to the development of any new technology. The elderly, the invalid, the blind, the estimated 45 percent of disabled Americans who work would all benefit immensely from self-driving cars. Individuals with drunk-driving convictions would no longer have to risk losing their jobs because they had no way to get to them. (There might not *be* drunk driving, period.) Long-distance truck drivers would no longer have to risk falling asleep at the wheel, endangering themselves and all around them. There might not *be* any long-distance truck drivers anymore, either—or cab drivers, or bus drivers, or delivery drivers—although surely there would be more of a need than ever for road crews to reach motorists stranded by breakdowns.

Self-driving cars are likely to change the entire culture of automobiles. It may be that almost no one will own a car at all. Instead, you'll order one up with your smartphone and leave it when you're done. The car will hurry off to its next assignment, instead of spend-ing 95 percent of its existence parked somewhere, as most cars do today.

Surely, many of us will miss the enjoyment of handling a superb piece of machinery on a highway. But computer-directed cars could move on highways at speeds impossible for humans to control; picture a driving experience like something out of *Mad Max*. You could spend journeys of thousands of miles working, reading, sleep-ing, even exercising in your own private compartment. Google is already working on a car that lacks a steering wheel or pedals.

In the future, some of us may or may not shoot up an ele-vator into outer space or hover in the air above electromagnets. But we will all be driven about like aristocrats in our computer-chauffeured autos.

Already on the road, self-driving cars, such as this one from Google, are likely to be ubiquitous within another ten years.

THE TRANSCONTINENTAL RAILROAD

E very one of the qualities that best exemplifies American invention would go into the building of the Transcontinental Railroad: vision, genius, private initiative, government backing, the courage of immigrant workers, and the participation of all. There was also plenty of greed, skullduggery, and self-dealing—but they would be overcome.

From 1830 on, Americans were broaching the notion that the railroad—which barely existed—could be flung across the breadth of the continent. Soon the idea was being promoted by one Asa Whitney, a distant cousin of Eli, the inventor of the cotton gin (see page 205) and a dry goods merchant in New York, who insisted that it would "place us in the centre of the world, compelling Europe on one side and Asia and Africa on the other to pass through us." In America, such a man could have his say as much as anyone else, and Whitney's vision lit a fire.

It was a fantastical idea, but only a natural extension of those grand projects, the Erie Canal and the transatlantic cable (see pages 9 and 71), that were putting the East Coast at the center of the Atlantic world. But Congress deadlocked on whether to build a "southern route" for the railroad or a "central route" from St. Louis to San Francisco.

The central route was cheaper and ran through more promising land, but it had one enormous impediment: the double-ridged granite walls of the Sierra Nevada, which climbed to seven thousand feet in just twenty miles, its river canyons too steep for the two-hundred-feet-per-mile rise that was the best any locomotive of the time could manage.

They weren't high enough to stop the engineer who built the first railroad west of the Rockies, Theodore "Crazy" Judah, so nicknamed for his monomania about finding a pass through the mountains on a grade a locomotive could manage. After four years of searching in the wilderness, Judah found one, the notorious Donner Pass, and started plying San Francisco, Sacramento, and Washington for backers.

In Washington, Judah found President Abraham Lincoln, who was not too busy winning the Civil War to neglect his own sweeping vision of the country's future. The

Pacific Railroad Act of 1862 was a complement to his Homestead Act of the same year, which virtually gave the country's vast western lands to those who would work them. The Railroad Act picked two companies, one starting from Council Bluffs, Iowa, and working west, the other starting in California and working east, to build the transcontinental road, backing them with land grants and government bonds.

In California, this meant the legendary "Big Four" storekeepers, Collis P. Huntington, Leland Stanford, Mark Hopkins, and Charles Crocker, who would do much to build modern California—at a hefty profit. Forming the Central Pacific Railroad with investments of a mere $1,500 apiece, they built a wagon road, while demanding greater subsidies and kicking Judah off the governing board when he demanded action. Crazy Judah stormed back east, determined to find backers to buy them out and, in a brutal irony, contracted yellow fever while crossing the Isthmus of Panama and died in his wife's arms at the age of just thirty-seven.

The venality on the other end of the road was even worse, an orgy of financial perfidy by the newly minted Union Pacific Railroad that set off repeated Wall Street panics and corrupted many of the nation's leading politicians. By the end of 1865, the Union Pacific had laid only forty miles of track.

Once they did start to move, though, both companies performed prodigious feats of engineering, working without anything resembling modern construction or earth-moving equipment. Cutting through the Sierra Nevada was done mostly with picks, shovels, wheelbarrows, one-horse dump carts, and hand drills. Central Pacific workers carved fifteen tunnels through the granite, with progress slowing at times to just two or three inches a day. Most of the Irish immigrant workforce drifted away, lured by hopes of finding gold or silver.

They were replaced by ten thousand Chinese. Despised by white Californians, subjected to special "permission" taxes and regularly attacked and lynched, the "Celestials" were usually paid about half the amount of white laborers on the railroad and had much of the rest of their pay seized by the Chinese "trading companies" that had smuggled them into the United States in the first place. They averaged just four-foot-ten and 120 pounds, but they proved to be wiry, tireless men who worked carefully and well. They were not afraid to work with a volatile new explosive, nitroglycerine, lowering each other down rock walls in gigantic baskets they wove themselves, drilling holes, pouring in the nitro, then calmly setting it off with a slow match. The building of the road was the process by which they became Americans. Before it was over, they had stood up to their bosses, staging a strike to win higher wages, and many would stay on in the land that had treated them so harshly.

Getting across the Great Plains was less dramatic but nearly as arduous, a project that involved hauling virtually all the fuel, food, water, and materials needed for ten thousand men across the prairie in transitory "hell on wheels" tent cities. Chief engineer Grenville Dodge and construction boss Jack Casement, a pair of tough Civil War generals, devised

The Transcontinental Railroad, completed in 1869, runs over a high trestle, through a tunnel, and under miles of sheds built in the Sierras to keep debris and snow off the tracks.

As soon as the "Golden Spike" was driven in, a telegraph message reading only "DONE" was sent to both coasts, setting off a national celebration.

The longest tunnel through the Sierras had to be dug through 1,659 feet of solid rock at the Number 6 Summit on the Donner Pass, at an elevation of 7,554 feet above sea level. Even with nitroglycerine, it took almost two years to complete.

Eventually, 80 percent of the 13,500 workers on the Central Pacific side of the job were Chinese. An estimated 1,500 died cutting their way through the Sierras.

By 1876, it took just eighty-three hours and thirty-nine minutes to get from New York City to San Francisco by the Transcontinental Express—only eleven hours and thirty minutes longer than it takes passengers on Amtrak today.

Between state and federal land grants, the Union Pacific and the Central Pacific received a total of 279,693,500 square miles of land.

The total investment of capital needed to build the railroad was over $100 million in 1860 dollars, or at least $3 billion in today's money.

twenty-two-car work trains, providing everything from a canteen to bunk beds to a machine shop and a forge for the task. Workers on the Union Pacific filled out the melting pot of Americans building the railroad, with freed black slaves, Civil War veterans from both sides, thousands of Mormons, and more Irish immigrants.

Both sides raced to get to the designated meeting point at Promontory Summit, Utah, with the Union Pacific laying seven miles of track in a day and Central Pacific's Chinese laying ten. The celebratory driving of the last "Golden Spike" on May 10, 1869, was as haphazard as everything else involved in building the Transcontinental Railroad—presidents of both companies missed on their first tries—but was a quintessentially *American* accomplishment, done in a headlong fashion, sometimes brilliantly, sometimes venally, but with a vigor and an audacity that would amaze the world.

The railroad companies would pay off the government with interest. They would haul mail and military personnel for free, bring farmers' goods to market and new immigrants to the West, and tie the country together. They would also serve our trade with the reborn nations of Asia, just as Asa Whitney and others had foreseen. The American rail freight system they pioneered remains the best and busiest in the world, pulling the wealth of nations over the same route that Crazy Judah discovered.

THE HUDSON AND EAST RIVER TUNNELS

T he colossal accomplishment that was the Transcontinental Railroad (see page 48) connected the entire country by rail in 1869. Except not quite. By the onset of the twentieth century, there was still no rail connection across New York's East River to Long Island, and still none directly over the Hudson and into Manhattan.

This state of affairs badly irked Alexander Cassatt, a gentleman horse breeder, yachtsman, and president of the Pennsylvania Railroad (and brother of the American painter Mary Cassatt). The "Pennsy" liked to call itself the "Standard Railroad of the World," but somehow its wheels still could not turn in New York City. A rail bridge across the Hudson would have to be twice the size of the Brooklyn Bridge and would cost an estimated $100 million. Building a tunnel below the Hudson had led to a grisly disaster in 1880 that killed twenty men. Running steam-powered locomotives through tunnels for that long a distance held grave risks in any case.

By 1901, though, electric traction motors had revolutionized underground rail-roading (see page 21), and Cassatt boldly decided to close *both* gaps—to build a series of tunnels that ran from New Jersey under the Hudson, then straight on through Manhattan, *and* under the East River to Long Island. He turned to New York's intrepid "sandhogs," the veteran construction workers who had been burrowing through every sort of mud, rock, and dirt since they'd started working on the Brooklyn Bridge in 1872—and who are still doing so today. They labored under the most difficult and dangerous conditions conceivable and were one of the very few integrated workforces in America at the time: African Americans, Irish, Italians, and others, all together in the brotherhood beneath the ground.

For the Hudson River tunnels, they had to blast and drill their way through 5,940 feet of the hardest traprock, from the Bergen Hill, New Jersey, tunnel portals to Weehawken—work that required constant and dangerous dynamite blasting. For every ten cubic yards of rock removed between Bergen Hill and Weehawken, work-ers used a foot of hardened drill steel and almost thirty pounds of dynamite. One hundred men, working on each of two ten-hour shifts, could still make only two to seven feet of progress a day.

At the same time, men were working from the bottom of the Weehawken shaft, seventy-six feet below ground, to build the two single-track tunnels under the Hudson. A small army of workers swarmed around the shaft entrance, manning machine and blacksmith shops (where they would repair, sharpen, and even redesign tunnel-building tools), an engine room, and a boiler room with three five-hundred-horsepower boilers, fueling the deafening air compressors that kept pushing in pressurized air and preventing cave-ins. Utilizing the new "shield method" of construction, shifts advanced behind a pair of massive iron cylinders. The men would literally push and jack the cylinders through the riverbed silt, squeezing the mud through the cylinder's apertures like sausage, then shoveling it up. Behind them, more men flanged together cast-iron and steel shells, twenty-three feet in diameter, covering them in two feet of reinforced concrete, building the tunnel walls as they went.

A similar operation was proceeding under the Hudson from the West Side of Manhattan. Here the sandhogs blasted their way fifty-eight feet down, through Fordham gneiss, the bedrock of skyscrapers, before pushing their own shield forward. Theirs was at the same time a more delicate operation, as they had to avoid unsettling foundations or rupturing the water or gas mains under the roughly five blocks they had to traverse between the front of the planned Pennsylvania Station (see page 57) and the river. But on September 10, 1906, a year *ahead* of schedule, the two shields of the north tube were halted within ten feet of each other to make sure they aligned. They did, to within one-sixteenth of an inch, in the 14,575-foot tunnel. "The shields met, coming together rim to rim, like two gargantuan tumblers," historian Lorraine B. Diehl would write. The New York sandhogs celebrated by passing a box of cigars through their cylinder to their New Jersey brethren.

Much work remained to be done. The East River tunnels were much shorter but more complicated, passing alternately through gravel and boulders, then the solid rock of a glacial ridge. Plagued by strikes, mistakes, and difficulties with maintaining air pressure, the East River project saw two men killed in a blowout that shot forty feet of water in the air. Dynamite that was set off prematurely in Long Island City killed three more. Once they were out from under the river, more problems awaited. Not only did these sandhogs, too, have to navigate the complicated underworld of Manhattan, from First Avenue to Seventh, dodging pipes and mains, but they also discovered a long-covered stream by Kips Bay that everyone had forgotten about—leaving the workers soaked with water. More seriously, a loose boulder killed two more men, and the constant dynamite blasting left a number of citizens seriously injured, many windows smashed, and everyone's nerves frayed for months.

Cassatt had gone to extraordinary lengths to look after his men, including onsite doctors and clinics, the provision of all the hot coffee they could drink, and heated lockers that dried their wet clothes—measures thought to help against "the bends," the painful and often deadly change in blood composition that came from working too far under the earth. Even so, twelve died in all from the bends, blowouts, dynamite accidents, and other mishaps.

The East River tunnels were finally finished in January 1908, the crews there passing to each other a toy train and a rag doll through the cylinders—the first "train" and the first "lady" to pass through the tunnel. That May, the entire magnificent series of tunnels was completed, connecting the entire country by rail once and for all.

But there was still one sticky problem: the tunnels under the Hudson were moving.

General Charles W. Raymond, chairman of the Pennsy's board of engineers, had reported the problem in the spring of 1906. It vexed Cassatt, who ordered vice president Samuel Rea and chief engineer Charles Jacobs to investigate. By 1907, Jacobs, measuring the movement of the tunnels with recording gauges, had the stunning answer: the tunnels were rising and falling with the tides.

Raymond felt the railroad should deal with the problem by drilling immense metal screw piles through the tunnels and down to the bedrock—something that had been planned from the beginning of construction. Rea, who had also studied the tunnels for two years, feared that screwing the tunnels to the bedrock might rupture them in the tides. Cassatt would have made the decision—but worn down by heart disease, worry over his grand project, and the flukish death of a beloved daughter, he had died in December 1906.

The final decision fell to Rea—who decided not to use the piles but to let the tunnels rise and fall with the tides. Over a hundred years of use have proved him right.

THE GENIUS DETAILS

Crews of twenty-four men worked at each shield in three eight-hour shifts. The average rate of progress in each heading was about eighteen feet per day.

The portions of the tunnels that ran under the Hudson River were 6,100 feet long. They required the excavation of nearly 190,000 cubic yards of material, 67,000 tons of iron and steel, and 57,000 cubic yards of concrete.

The Manhattan & Hudson Railroad reopened and completed a smaller, previously failed tunnel for a commuter line between New Jersey and New York, before the Pennsy completed its tunnels in 1908. This eventually became the Port Authority Trans-Hudson (PATH) train.

THE HUDSON AND EAST RIVER TUNNELS

AMERICA'S TRAIN STATIONS

T o crown the magnificent tunnels that linked Manhattan and Long Island to the rest of America (see page 53), Alexander Cassatt, president of the Pennsylvania Railroad, hired the leading American architectural firm of the era, McKim, Mead & White. Charles McKim's Pennsylvania Station would become a legend, a Beaux-Arts masterpiece "vast enough to hold the sound of time," an awed Thomas Wolfe would write.

Its inspiration? "About 25 centuries of classical culture and the standards of style, elegance and grandeur that it gave to the dreams and constructions of Western man," was how the architectural critic Ada Louise Huxtable would describe it.

The main hall, with its vaulted, coffered ceiling that had so astonished Wolfe, was taken from the ancient Roman Baths of Caracalla, but other influences were plucked from here and there, up and down the centuries: the thirty-five-foot-high Doric columns of the station's exterior, inspired by the colonnade Bernini had used to enclose the Piazza di San Pietro at the Vatican in the seventeenth century. The steel-and-glass roofs, and train concourses, from the new Gare du Quai d'Orsay in Paris. The porticoed carriageways inspired by Berlin's Brandenburg Gate. The shops and waiting rooms were clad in pink Milton granite and honeyed Travertine marble—the same stone used to build Hadrian's Tomb, the Roman Coliseum, and the Basilica of St. Peter's over the ages, a rare marble that glows more beautifully the more it is rubbed by the passing humanity. These profound spaces were lit through lunette windows, held up by Corinthian and Ionic columns and festooned with statuary, huge clocks, and enormous wall maps of the great nation through which the Pennsylvania Railroad roamed.

It was the work of a nation bold and confident enough to borrow the creations of a dozen other civilizations, and to imbue them with a purpose and an energy all its own.

Just across town—in the same city, and the same decade, it would inspire in turn something perhaps even more lustrous. If Penn Station was vast enough to hold the sound of time, Grand Central Terminal was where time began, where America's first official clock was installed once the country was first divided into four official time zones—times imposed on a continent by a people who now traversed it more quickly than the progress of the sun.

The front of Cincinnati's Union Terminal today.

If it was smaller than Penn Station, Grand Central was a good deal more passenger-friendly, its adornments at least as beautiful with its friezes of winged train wheels, statues of Greek gods, electric chandeliers, the famous opal clock in the middle of its main room, and the breathtaking cerulean blue celestial map on the 120-foot-high ceiling of its Grand Concourse.

Uses built upon uses. Set in the heart of Manhattan, Grand Central was a frenetic transportation hub, a "monument to movement." When it was built, it was a central node not just of intercity rail lines but also of subway, trolley, and elevated rail lines, soon to be bisected by cars running right through it—wheels within wheels. Park Avenue itself was an invention of the terminal; once an open train ditch, it was covered over when the trains were converted to electricity and transformed almost instantly into one of the wealthiest and most coveted addresses in America. Around the great edifice would spring up "Terminal City," its own cluster of major hotels, office buildings, and residences, spurring a building spree in the area that would come to include the Chrysler Building, probably the most exquisite skyscraper ever constructed.

Pennsylvania Station and Grand Central Terminal were the most magnificent buildings ever constructed to that point without reference to king or church, built *and* used by the free citizens of a free country. And they remain vital nodes of travel and commerce. Nearly one hundred million commuter train passengers a year today use Grand Central Terminal, a new all-time high. The old Pennsylvania Station served a record 109 million passengers in 1945, or almost 65 percent of all intercity rail travel at the time. Today, an estimated five hundred thousand commuter and intercity passengers a day, or 182.5 million a year, use Penn Station, making it the busiest transportation hub in North America, with a higher traffic share than New York's three regional airports combined.

Yet they were not unique. Visit almost any midsize city or even town in America between the wars, and you were as likely as not to see a masterpiece when you stepped off the train. Forth Worth, Texas, a city of less than 200,000 before World War II, boasted not one but two brilliant showcases, the late-Victorian, neoclassical Santa Fe Station and the towering art deco Texas and Pacific Station. San Antonio, with just over a quarter million souls by 1940, had *three*, the Katy, Mopac, and Sunset stations, each of them a Mission-style gem.

Nearly every American architect worth his salt tried his hand at a train station. Most of what they produced were Beaux-Arts beauties, but they encompassed pretty much every known style of design. Even in the smallest towns, train stations tended to be little jewels of craftsmanship: the charming Rhinecliff Station in New York, built cunningly up the side of a cliff along the Hudson River; Durand Union Station in Michigan, with its terrazzo floor, marble wainscoting, and Château Revival turrets; Kentucky's Colonial Revival beauty in Maysville. They reflected the care and pride of people who believed they were somebody, no matter where they might happen to be.

"Each station functioned as did the main gate on a medieval town: it was the welcoming portal to that community, and it was meant to impress, comfort, and reassure the visitor . . . ," wrote Ed Breslin and Hugh Van Dusen in *America's Great Railroad Stations*. "All embodied America's love of, and genius for, commercial excellence. Whether built on the scale of a small chapel, a substantial church, or a monumental cathedral, all of these stations personified and reflected America's secular spirituality fueled by the belief that life could endlessly be enhanced by aesthetic beauty, industrial might, technological know-how, and creature comforts while traveling in style with alacrity from one point on the compass to another."

The end of the fifty years or so of the great station-building era would come in the late 1930s with a grand finale: Los Angeles's coolly gorgeous Union Station, a combination of art deco, Mission Revival, and Spanish Colonial styles that seemed as if it had been built as a set for a hundred *films noir* (which, in the end, it was), and Cincinnati's Union Terminal, completed in 1933, which really doesn't look quite like anything else ever built in this country. Its marble and concrete facade resembles a sort of gigantic art deco radio, with a clock set in the middle. Inside, its main rotunda flowed with sunlight in the daytime, and glowed at night, and nestled in its apex was the largest semidome ever built in the Western Hemisphere, crowned by a brilliant aquamarine half-ocular glass and surrounded by ever-widening rings of color, as if someone had brought the planet Saturn down and tied it to Cincinnati.

Every need of the traveler was anticipated, right down to toy shops, baths and showers, and a newsreel theater, and everywhere one looked, there were murals showing the rise of the city and the progress of the nation. Every surface one's eye or hands or shoes came in contact with—from the Carrara glass toilet partitions in the bathrooms to the cork floor tiles to the hand-cut wallpaper to the leather-cushioned seats and zebrawood veneers—was something new and elegant and enthralling. And why shouldn't it have been? For it was built for the people.

The total area of the Cincinnati Union Terminal was 287 acres, including what would be lavish gardens, fountains, and curved drives for cars, taxicabs, buses, and trolley lines to reach the terminal and turn around.

F. Winold Reiss created two massive paintings for Union Terminal that were turned into two glass color mosaics, each 22 feet high by 110 feet long. They depict the history both of Cincinnati and of American transportation.

Pierre Bourdelle's flourishes in Union Terminal included his mermaid murals, on either side of the newsreel theater screen; aluminum structured seats with soft leather cushions; a bookshop shaped like a train observation car; a men's lounge with railroad motifs and "zebrawood Flexwood veneer"; and depictions of the sun, the moon, and the planets dangling from the ceiling of the toy shop, among many others.

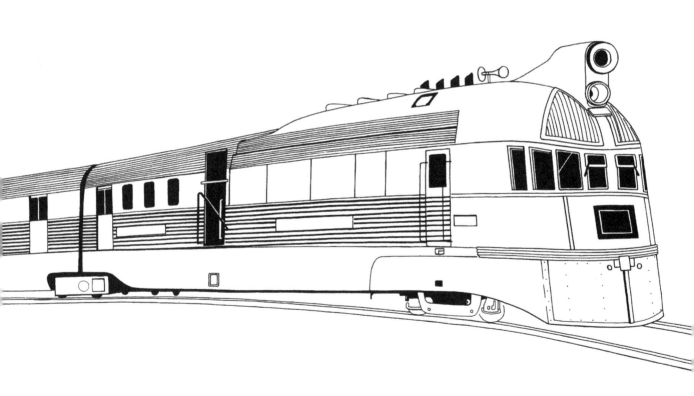

STREAMLINED TRAINS

I t came storming down from Denver, out of the Rockies and across the Plains one morning in the late spring of 1934. Named for the west wind, it made it all the way to Chicago before dark, "1,015 miles in 785 minutes," and as it went the crowds began to gather to watch it pass, drawn by news reports on the radio. By the time it rolled onto a huge stage, as the grand finale of the "Wings of a Century" pageant at Chicago's Century of Progress Exposition, the people waiting to see it could no longer contain themselves and poured down out of their seats just to touch it.

The train was the Burlington Zephyr, and it was something brand-new in train travel, a perfect confluence of bold design and engineering. No train had ever gone so fast before. None had ever traveled so lightly. None had ever gone so far without having to stop for fuel or water. No train ever had or ever would so capture the imagination of the American people.

The Zephyr's "Dawn to Dusk" run cut the time of the regular, steam-driven passenger train from Denver to Chicago in half. It arrived an hour ahead of its own unparalleled schedule, reached a top speed of 112 miles an hour, and averaged nearly 78 miles an hour—faster than any American train travels today. With diesel fuel at five cents a gallon, it had done all this for just $14.64.

The Zephyr marked a particularly happy collaboration of some of the most innovative minds in America, but it was only the vanguard of a revolution. Its extraordinarily new, light, and efficient diesel engine was the brainchild of Charles Kettering, the engineering genius who had developed the automatic transmission for automobiles. Edward G. Budd, the metallurgist who built the Zephyr's cars, had been working for almost twenty years building automobiles out of a particularly light but durable version of stainless steel, a European invention that had emerged shortly before World War I. With one of his mechanical engineers, Col. Earl J. W. Ragsdale, Budd built his trains with low centers of gravity so they could take curves at speed and built them sleek to reduce wind resistance. He had also figured out how to "shot-weld" stainless steel. This was a groundbreaking process that consisted of welding two pieces of stainless steel together by passing a high electrical current through them for a short time. With twice the shear strength of a rivet weld, shot-welding also preserved the

The Burlington Zephyr drew trains into a new era of beauty, speed, and efficiency.

The first streamliner train, beating the Burlington Zephyr out of the yard by two months, was the Union Pacific's M-10,000.

Also a revolutionary design, the M-10,000 had three cars that together weighed only eighty-five tons; a traditional ten-car steam-powered train weighed one thousand tons.

In 1919 Ralph Budd became the youngest head of an American railroad at age forty, when he took over the Great Northern Railroad. He oversaw the cutting of the 7.8-mile Cascade Tunnel through the Cascade Range in central Washington. It was the longest rail tunnel in America at the time and is still one of the longest in the world.

By the late 1930s, the Zephyr's design, as the PBS documentary *Streamliners* noted, "could be found on everything from toasters to tractors and corsets to coffins."

The streamliners' power, speed, and efficiency proved to be a tremendous advantage to the United States during World War II, with trains carrying 97 percent of all troops and 90 percent of all military freight.

strength and rust resistance of stainless steel, an ideal metal for trains: strong, light, malleable, and attractive, it does not rust, despite exposure to the elements.

Bringing all this brilliance together was Ralph Budd, a very distant relation to Edward, and a veteran railroad man who had left his Iowa farm, gone through high school and college in just six years, and worked for the great engineer John F. Stevens in building the Panama Canal (see page 16). After running the Great Northern Railroad through the 1920s, he took over the Chicago, Burlington & Quincy on the first day of 1932.

His timing could not have been worse. The Burlington was a tired, crumbling line that had lost 60 percent of its passengers since 1926.

But the major railroads were beginning their long conversion from steam to cheaper diesel, or even electric power. They tested their engines and cars in wind tunnels, dropping fuel consumption by 40 percent and ratcheting up top speeds. They hired such leading industrial designers as Raymond Loewy, Norman Bel Geddes, Otto Kuhler, William B. Stout, Ralph H. Upson, and Henry Dreyfuss—incredibly versatile artists whose work in many fields was changing the whole "look" of America.

This was usually to make the new trains *look* new, in art deco, or art deco's late stage, Art Moderne or Streamline Moderne—French ideas that Americans would take to phenomenal new heights. It was a style that adapted beautifully to the immense scale of the machine age.

"Brute force can have a very sophisticated appearance, almost of a great finesse, and at the same time be a monster of power," wrote Loewy of the sleek, stylish CGI locomotive, capable of 8,000 horsepower, that he helped design.

It was true. Their fluted, stainless steel skins made the trains seem *fast*. The Zephyr engines sported cowls like the helmets of ancient Greek warriors. Dreyfuss and Loewy's bottle- and shark-nosed locomotives and Bel Geddes's serpentine creations look futuristic to this day. Engines and cars alike were now streaked with brilliant, supergraphic bands of black, maroon, light gray, orange, red. They were painted in bright yellow and brown, blue and gray; lettered in gold; decorated with chrome fins. The Seaboard Air Line trains, from Richmond, Virginia, to Miami, looked like a spectacular tropical sunset on wheels. The name "Air Line" on a train was no coincidence. The railroads drew on the most popular designs of their rivals, planes and cars and even dirigibles, which borrowed freely from them in turn.

Inside their passenger cars—and in many of the ticket offices and the stations that served the new trains (see page 57)—was an amazing consistency and sophistication of design. Travelers were treated to the newest, most appealing materials and decorations, including Montel metal, Bakelite, cork, polished aluminum, Formica tops, steel seats and tables, inset ceilings, sans serif lettering, and silk drapes. The streamliner trains set new standards in comfort, with roomy observation cars, air-conditioning, soft lighting, and cushioned seats replacing the old red-plush chairs that previous passengers had stuck to on hot days. They offered some of the best public dining anywhere in America, available at almost any time of the day or night, served on real china with glassware, silverware, and fine linen napkins.

"Although galleys had barely room to flip a pancake, the cooks turned out an extraordinary range of dishes," wrote historian Bill Bryson. "On Union Pacific trains, for breakfast alone the discerning guest could choose among nearly forty dishes—sirloin or porterhouse steak, veal cutlet, mutton chop, wheatcakes, broiled salt mackerel, half a spring chicken, creamed potatoes, cornbread, bacon, ham, link or patty sausage, and eggs in any style—and the rest of the meals of the day were just as commodious."

The public came flocking back; the Zephyrs alone sparked a 26 percent increase in passenger travel. Millions of Americans paid to see the sparkling new trains even before they came on board, crowding the Chicago and New York world fairs and train stations all over the country, where Ralph Budd and his fellow rail executives had smartly put them on display. Though their heyday would glimmer only a little longer than those other fantastically beautiful creations, the Yankee clipper ships (see page 13), the streamliners would leave an indelible impression on all who rode or saw them.

THE ROTARY PRINTING PRESS

E verywhere along the city streets of the young republic, men and women could be seen reading in public, subscription, and society libraries. People of all classes, backgrounds, and professions, reading in their every spare moment. Impressed deep in the psyche of America, the most literate society ever founded, was the idea that you could improve yourself—morally, philosophically, *financially*—through the written word.

America's publishers struggled to keep up. While the Revolution had been fomented and sustained in large part by committees of correspondence, printing was still little more advanced than it had been nearly four centuries before, when that Rhenish goldsmith Johannes Gutenberg started hand-cranking Bibles.

A big change came early in the nineteenth century, when the printing press, like everything else, was run by steam power. English and German printers began building cylinder steam presses that greatly increased the speed of printing. But it was still not fast enough for the American reading public. By the 1840s, most cities boasted at least a dozen daily newspapers, many of them priced to sell at a penny a paper. These "penny dreadfuls," as they were often called, battened on terrible or thrilling events. As news moved faster than ever, thanks to the new telegraph (see page 67), they printed "extra" after extra edition to keep readers apprised of the very latest turn of events in battles, murder trials, famous deathbeds, elections, and out-and-out hoaxes.

Speed was more required than ever, and it was achieved by one Richard Hoe, son of an English immigrant and printer. Richard had gone into his father's business at fifteen and inherited it at twenty-one. Like his father—like so many acolytes of the esoteric, now largely vanished profession of newspaper printing—he was always looking for ways to improve the process.

What he did was discard the traditional flatbed press on which newspapers were printed and make the type revolve instead. Hoe placed large rotating drums on the fixed platform of his press—something inventors before him had considered, only to be thwarted by how to get the "stereotype," the flat plate on which the type was set, to stick on a curved drum. Hoe nearly despaired over the same problem. But working at it all through the night, he came up with an idea that would become known as "the turtle": a curved *mold* of the stereotype that would fit into a curved box fitted exactly to the drums.

Made originally out of plaster of paris but soon out of papier-mâché, the turtle shot printing into the modern age. The Hoe rotary or "lightning press" was first used by a particularly lurid penny dreadful, the *Philadelphia Public Ledger*, where it proved an instant success, printing eight thousand sheets of paper (on one side) in the space of an hour—far faster than any press then in existence.

Hoe would go on to make 175 modifications to his original model, turning it exponentially faster and more efficient. He invented a highly successful grinding saw, to better cut the newspapers on his presses, and founded a saw company. Its factory even included free instruction for apprentices.

Yet the next great step forward was taken by William Bullock, an orphan from upstate New York who became a printer and inventor and eventually the father of thirteen children. Understandably eager to maximize his earnings, Bullock invented a crucial automatic feed of the vast rolls of paper—sometimes as much as five miles long—that

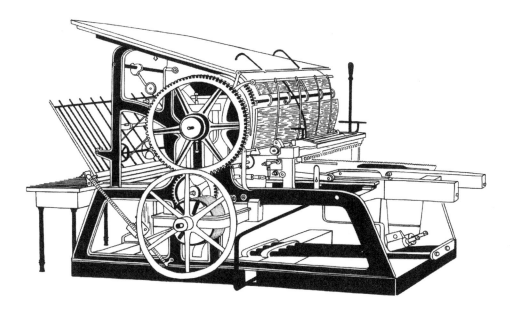

The Hoe rotary "lightning" press.

were fed into urban presses, thereby eliminating the backbreaking work of hand-feeding. His press also printed on both sides; cut each copy with an automatic, serrated knife; folded the paper; adjusted itself; and generally did everything save tuck the operator into bed at night. Bullock got his press up to thirty thousand sheets an hour. But one day at the *Public Ledger*, in a macabre accident, Bullock tried to give the press a "Brogan adjustment," kicking a driving belt back into a pulley. His leg was caught and crushed, it developed gangrene, and he died on the operating table.

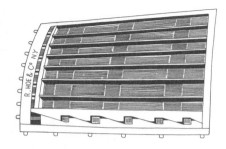

The "turtle": the key to the Hoe lightning press was molded out of plaster of paris, so it could be curved to fit the drums.

Hoe, meanwhile, would continue perfecting his presses until, by 1882, he produced an almost completely automated machine. With other inventions, such as linotype, newspapers would get steadily bigger, cleaner, and more easily readable, stuffed with Sunday supplements, and—after the invention of a color process—even comics, for America's growing middle class. It had never been easier to read all about it.

THE ELECTRIC TELEGRAPH

F or all the inventions roiling the Western world in the first half of the nineteenth century, none caught the imagination of Americans like the telegraph. It was the telegraph that first destroyed traditional notions of space and time and ushered in modernity. And we owe it all to the efforts of a bombastic bigot, a corrupt congressman, and a former child actor.

The theory of an electric telegraph had been bandied about since the mid-1700s, and by the early 1800s an array of inventors throughout Europe and the United States were creating progressively more sophisticated systems of conveying information by means of electricity. William Fothergill Cooke and Charles Wheatstone in England had even developed a commercial, rail-linked telegraphic system by 1837. Yet all of these systems were hugely cumbersome and limited, passing along electricity through galvanic fluids or requiring multiple wires to send messages by way of needles that pointed, one by one, to letters arrayed around a clockface.

Samuel Finley Breese Morse, a native of Charlestown, Massachusetts, was certain that he had a better idea, aided by the fact that he paid almost no attention to what anyone else in the field was doing. A windy, priggish racist, he believed fervently that God wanted slavery preserved, ran twice for mayor of New York on the anti-immigrant, anti-Catholic Know-Nothing ticket, and led a campaign against "French dancing" in theaters.

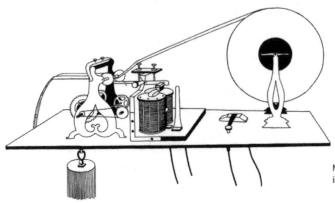

Morse's telegraph, a year after its invention and complete with ticker tape.

Yet Morse was also blessed with immense talents, ingenuity, and determination. An amateur inventor from an early age, a portraitist of almost the first rank, Morse was working on a painting of Lafayette in Washington, D.C., in 1825, when his wife suddenly took ill. She was dead before he could be notified and buried before he could get home, a loss that haunted him.

There had to be a way to convey news more quickly. While teaching at New York University, Morse toiled for years on making an electric telegraph but still could not produce a signal that traveled more than twenty feet.

He needed help—and he got it. A fellow congregant at his church named Alfred Vail helped him develop the "Morse code" that would replace Morse's own unwieldy idea to assign every word in the dictionary a number. Instead, the code that Vail and Morse worked out was a sort of binary system, which is one reason why many historians refer to the telegraph as the "Victorian Internet." The "keys" that telegraph operators tapped transmitted series of what were either dots and dashes, with a dot (or "dah") created by taking one's finger off the key quickly, and a dash (or "dit") by leaving the finger down a little longer.

This made Morse's telegraph run much faster than any existing wire system. Vail also talked Morse into replacing his telegraph's registry pencil with a stylus that perforated paper, and the composing stick that sent messages with a "key," a piece of metal tapping on another piece of metal. Operators could now *hear* messages even before they saw their marks and thus could further speed transmission. Finally, Vail talked his father into investing $2,000 in Morse's telegraph and letting them produce a working model of it at his Speedwell Ironworks in New Jersey.

Around the same time, a fellow professor at NYU, Leonard Gale, vastly improved Morse's circuit and introduced him to Joseph Henry. Henry, a former child actor, was a brilliant polymath who would later advise President Lincoln on science and would run the Smithsonian for decades. While teaching at the Albany Academy, in Albany, New York, he had discovered the property of self-inductance, and built the much stronger electromagnet that would be necessary for telegraphy of any distance. In 1831—well before anyone else—Henry amused his students by inventing the first operational magnetic telegraph in the world, using his electromagnet and a primitive battery to send an electrical signal around a mile of the academy's walls until it made an armature strike a bell.

Henry, though, had no interest in anything beyond providing his scientific discoveries to others, and he left it to Morse, Vail, and Gale to work out a commercial telegraphic system.

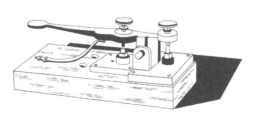

Morse's—and Alfred Vail's, and Leonard Gale's—original telegraph key, which conveyed the Biblical verse "What hath God wrought!" from Washington to Baltimore in 1844.

Morse's greatest contribution was figuring out how to use his strengthened electromagnets to continually relay the electrical charge through the wires.

"If I can succeed in working a magnet ten miles," became Morse's mantra, "I can go around the globe."

He and Vail got it to go ten miles at Speedwell and began lobbying Congress for funds, bribing Rep. Francis O. J. "Fog" Smith with a one-fourth share in Morse's enterprise. Funds for the telegraph passed the House of Representative by a margin of six votes. But as midnight approached on the last night of the March 1843 congressional session, the Senate still had not taken it up, and there were 140 bills on the calendar ahead of it. When midnight came, the session would end, and the whole effort would have to be taken up months later—if it ever was.

Morse, who had watched anxiously for days from the galleries, returned to his boardinghouse in despair. The next morning, he was amazed to learn his funds had been gaveled through with five minutes to spare.

Morse now had $30,000 to build a thirty-eight-mile demonstration line from Washington to Baltimore. He hired as contractor an acquaintance of Fog Smith's named Ezra Cornell. Cornell was supposed to bury Morse's wires in the ground, insulated in pipes, but they kept shorting out. With time and money running out, Cornell convinced Morse to let him string the telegraph wires between poles along the Baltimore & Ohio Railroad's right-of-way and insulate them with glass drawer knobs.

On May 24, 1844, Morse telegraphed the fateful biblical verse to Vail in Baltimore and received it back: "What hath God wrought!"

He had invented the first practical electrical telegraph, and it would sweep the world. Morse's device had found its time—just as the railroads were stitching together the nation, clearing access lanes throughout the wilderness for its poles, and creating a need for a communications device to run with and ahead of the trains. The telegraph was "faster than the sun," shrinking the great distances of America and then the world more dramatically than anything ever had, or ever quite would again.

THE GENIUS DETAILS

Morse would offer his entire invention and its patents to the US Postal Service for $100,000—a fantastic bargain, as it turned out. The government turned him down.

The famous telegraphic plea for help, "SOS," does not actually stand for anything. Its combination of three "dots" for "S" and three "dashes" for "O" is just one of the easiest combinations to send and hear.

By 1849, there were over twelve thousand miles of telegraph wire in America, run by twenty different companies. In 1856, Hiram Sibley and Ezra Cornell consolidated most US telegraph companies into what would become Western Union. AT&T, which ran Bell Telephone, gained control of Western Union in 1908. Government antitrust efforts forced AT&T to give it up in 1913.

The telegraph was connected across America by 1861, bringing an end to the Pony Express.

The final Morse code signal was transmitted in the United States on July 12, 1999, repeating Morse's initial biblical verse.

"FASTER"

THE TRANSATLANTIC CABLE

By 1860, just sixteen years after the original thirty-eight miles of Morse's wires were strung up (see page 67), there were fifty thousand miles of telegraph lines in the United States, or about 40 percent of all the mileage in the world. Every year, some five million messages zipped back and forth between Americans—but it still took a ten-day ocean voyage to get any news from Europe.

Telegraph cables had recently been laid across the English Channel and New York Harbor, but deep ocean? Any cable would have to be incredibly well insulated and somehow avoid deepwater canyons and jagged rocks. Who knew if ships could even carry all the heavy coils of cable necessary, much less lower them smoothly enough that they did not snap or pile up on themselves? Who knew if telegraphic signals, without the relays available on land, could travel such a distance at all?

Cyrus Field was willing to find out. A paper magnate and art patron who, whenever he visited a foreign country, always asked first what the word for "faster" was, Field was introduced to a telegraph company owner with a scheme to lay an underwater cable across the Cabot Strait, from Newfoundland to Nova Scotia. Field proposed a more audacious idea: Why not lay a cable all the way across the Atlantic Ocean?

Living up to his favorite word, Field quickly raised $1.5 million in private funds. Consulting the country's leading oceanographer, Commander Matthew Fontaine Maury, he was informed that recent deep-sea soundings indicated there was a perfect "telegraph plateau" across the North Atlantic. When most of the company's seed money was exhausted in the unexpectedly difficult effort just to get the cable across Cabot Strait, Field rushed off to England on what would be the first of more than thirty transatlantic crossings for his baby. There British foreign secretary Lord Clarendon asked him, "Suppose you make the attempt and fail—your cable is lost at sea—then what will you do?"

"Charge it to profit and loss, and go to work to lay another," Field replied, so impressing Clarendon that he secured for him a subsidy of £1,400 a year and the promise of a ship.

Field raised still more money from London investors, then raced back to America to lobby Congress. After two months of heated debate, backing for the cable passed by

The USS *Niagara* and the HMS *Agamemnon* laying the first transatlantic cable in 1858.

Field's tombstone reads: "Cyrus West Field, To whose courage, energy and perseverance, the world owes the Atlantic telegraph."

The insulation of Field's first Atlantic cable was probably burned out by the two-thousand-volt shocks that the English surgeon Edward Orange Wildman Whitehouse convinced him to use in sending messages.

Field's first cable had three layers of gutta-percha wrapped around seven strands of copper wire. The second cable had four layers of gutta-percha and an adhesive known as "Chatterton's compound" between the layers. Polyethylene would eventually replace gutta-percha as the main insulation of underwater cables.

The first public telegraphic message across the Atlantic, in 1858, was just ninety-nine words from Queen Victoria to President James Buchanan, but it took sixteen and a half hours to transmit. Submarine cables today can transmit eighty-four billion words per second.

The initial rate in 1866 for messages was $10 a word, with a ten-word minimum.

one vote in the Senate. President Franklin Pierce signed the Atlantic Cable Act into law on his last day in office, and the transatlantic cable quickly captured the imagination of America. The *New York Herald* called it "the grandest work which has ever been attempted by the genius and enterprise of man."

The cable itself was manufactured in England, seven strands of copper wire connected strand by strand by riggers, with a copper penny soldered in for luck. All 2,500 nautical miles' worth were wrapped in gutta-percha, a sort of natural plastic extracted from the sap of Malaysian trees. Too heavy for any one ship, it was carried on both the USS *Niagara* and the HMS *Agamemnon*.

The Atlantic would not go quietly. The cable broke the first day, was repaired, then went dead for a few hours for no perceptible reason. Then, just four hundred miles from Ireland, it snapped after a heavy wave hit the *Niagara* and sank in water two miles deep. They tried again the next year. But the *Agamemnon* nearly sank in a storm, and the cables repeatedly snapped again.

Field stayed publicly calm, but privately his resolve had begun to crack. Only the intervention of his friend and president of the company, Peter Cooper, had kept his paper business from going bankrupt during the 1857 financial panic, and now the cable had failed again.

The *Niagara* and the *Agamemnon* set off once more. All the metal aboard the *Niagara* made its compass go awry and pulled it off course, but the problem was discovered in time and another ship sent ahead to show the way. The *Agamemnon* nearly ran out of coal and had to rely on its sails to get back to Ireland. But at 1:45 on the morning of August 5, 1858, Field rowed himself up on a Newfoundland beach and woke the local telegraph operators with the announcement "The cable is laid!"

Both sides of the Atlantic erupted in joy—prematurely. Messages across the Atlantic were faint and interminably slow. Within three weeks, they had stopped altogether. The House of Commons held its first committee of inquiry on a technological matter and found that the cable was essentially "blown out" from massive electrical

The machinery that laid the transatlantic cable from the deck of the *Great Eastern*, in 1866—the first cable that would remain in continuous use for more than a few weeks.

bursts. Field had proceeded without sufficient testing and preparation. "Faster" was too fast.

Field learned from the committee's findings. In July 1865 he was back at it again with more investors, better cable, and a single, gigantic vessel, the *Great Eastern*, the largest ship yet put afloat. As it proceeded across the Atlantic from Ireland, multiple flaws were found in the cable's connections, and each time it had to be hauled back up, as one reporter wrote, like "an elephant taking up a straw in its proboscis." The breaks were caused by small spikes in the line, a fault in the iron sheathing around the cable. Each flaw was corrected—but during one such procedure, less than six hundred miles off Newfoundland, the cable snapped and sank. Eleven days of grappling the ocean floor failed to retrieve it.

"We've learned a great deal, and next summer we'll lay the cable without a doubt," Field announced.

He was right. On July 27, 1866, the first transatlantic cable was successfully connected.

Many more cables would follow, and communications around the world—and particularly between New York and London—were soon so fast that it was another reason why historians referred to the telegraph as the "Victorian Internet." Phone and fiber-optic cables followed the telegraphic lines, and to this day they remain much cheaper than satellite phone connections—and vital to world finance. A privately owned cable line was installed in 2010 simply to reduce the "latency" of a call across the Atlantic from sixty-five to sixty milliseconds.

As for Field's old cables, most of them are still down on the ocean floor, including the very first, failed ones. Ships repairing new wiring occasionally pick them up by accident, a reminder of our first connections.

THE TELEPHONE

B y the mid-nineteenth century, the idea that human communications could move well beyond the dazzling new telegraph, that the human voice could even be conveyed over long distances, was in the air. A French telegraph engineer, Charles Bourseul, had proposed the basic idea of the telephone in 1854, and six years later a German teacher named Johann Philipp Reis developed a "make and break" "telephon," as he termed it.

Reis's invention proved able to transmit very faint and indistinct musical notes, other sounds, and even some speech in much the same way a telegraph works, by "making" and then "breaking" the current—something that also made an actual conversation impossible.

A young language teacher on the North Shore of Massachusetts—a Scottish immigrant whose family had immigrated to the United States via Canada to escape what seemed to be a curse of tuberculosis on it—was moving in a more promising direction.

Alexander Graham Bell was the sort of man whose idea of bedtime reading was scouring the *Encyclopaedia Britannica* for new ideas. A prodigy, he learned to play the piano as a child with no formal training, and at twelve he invented a machine to dehusk wheat at a flour mill. By sixteen he was not only learning Greek, Latin, elocution, and music but giving classes in these subjects as well. But all his life, Alexander Graham Bell would attempt to turn his work to the betterment of mankind.

Bell's family had long been experts in elocution, and Alexander himself taught speech to deaf students in Boston, including Helen Keller and fifteen-year-old

A Western Electric dial candlestick phone, from the 1920s. In 1892, the candlestick became the first upright, desktop phone.

Mabel Hubbard, who had lost her hearing to a bout of scarlet fever and who would eventually become his wife. Thanks to the sponsorship of his future father-in-law, Gardiner Hubbard, and the father of another pupil, Thomas Sanders, Bell was able to cut down on his teaching and concentrate on his experiments.

By May 1875, Bell thought he had "the germ of a great invention," a way to convey the human voice over the telegraph using undulating electrical currents via reeds tuned to different frequencies. When he confided to Joseph Henry, the genius director of the Smithsonian who had contributed so much to the invention of the telegraph (see page 67), that he still felt stymied by his lack of knowledge of electricity, Henry told him simply, "Get it!"

It was all the encouragement Bell needed, though much of that knowledge came in the form of Thomas A. Watson, a bright young electrical designer and mechanic whom Bell hired as his assistant. On June 2, 1875, Watson plucked one of the mobile metal reeds on their "acoustic telegraph" and transmitted the tone over the wire, making the reed on Bell's end vibrate. It was done without "breaking" the current. Indeed, the battery did not generate the transmission at all—it only provided the magnetic field through which the magnetic-electric currents conveyed the sound.

The next month, Bell and Watson designed and built a receiver with a diaphragm of stretched goldbeater's skin connected to a hinged armature. By the end of the year, Bell was ready to file patents in Great Britain and the United States, but he still had no real idea if his apparatus worked—he had not been able to produce anything more than indistinct "voice-like sounds" over the wires.

He received his patent on March 7, 1876—and three days later, working at their lab in Boston, Bell used his new "vibraphone" to call to Watson in another room—according to folklore, because he had spilled some acid: "Mr. Watson—come here—I want to see you." The words came through loud and clear.

It was the first time that "articulate speech," as Bell called it, had been transmitted intelligibly over a wire by means of electricity. After a few more experiments with their new device over distances of up to four miles, Bell and his backers offered to sell their patent outright to the president of Western Union for $100,000. He said no. Bell, an able salesman, took his telephone to the 1876 Centennial Exposition in Philadelphia, where it amazed visitors from around the world. Soon after, he married his Mabel (now twenty) and took her on an extended "working honeymoon" around Europe, demonstrating the phone for Queen Victoria, who deemed it "most extraordinary"—and within two years the head of Western Union was telling friends that if he could get the patent now for $25 *million* he would consider it a bargain.

This may have been apocryphal, as Bell and his associates still faced a stiff court fight. A number of other inventors had been working on similar devices, and these extended battles would eventually lead to the bribing of a US attorney general and all sorts of wild charges against Bell. None of this was vaguely true, and thanks to his copious record keeping and previous patent applications, Alexander Graham Bell was able to prove he had been first.

His telephone was still a rudimentary device that often required one to shout into it. Just as Bell would improve Edison's phonograph by introducing a version that played wax records, Edison returned the favor by working with Charles Batchelor to invent a carbon button transmitter for the telephone that spared callers' vocal cords and enabled transmission over greater distances. In January 1915, Alexander Graham Bell made the first transcontinental phone call—this time from Lower Manhattan to Thomas Watson in San Francisco. Their 3,400-mile call was actually clearer than their first one, though Watson remarked that it would take him a little longer to come to Bell this time.

Bell's sponsors, Hubbard and Sanders, would soon buy up Edison and Elisha Gray's similar patents and merge Bell Telephone with Western Union, creating the American Telephone and Telegraph Company. Meanwhile, though, Bell had moved on to a dozen new ways to improve the human condition. In 1880, he became the sixth man ever to be awarded the "Volta Prize," established by Napoleon in 1801 to honor the Italian inventor Alessandro Volta (see page 193). He selflessly used the $10,000 (about $250,000 in today's money) he won with the vaunted Volta Prize to establish the Volta Bureau, which promoted "the Teaching of Speech to the Deaf," and the Volta Laboratories, which supported

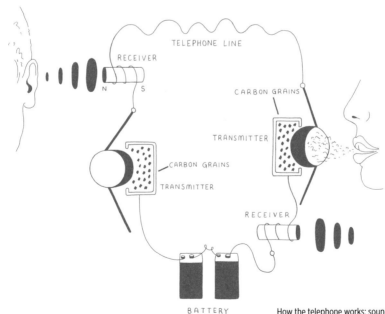

How the telephone works: sound is transmitted electrically, through wires, and then to a receiver tuned to different frequencies, made originally of reeds and later of stretched goldbeater's skin and then carbon.

promising experiments by himself but also other scientists, all of whom would keep the rights to their own patents.

The Volta Labs would inspire the creation of the famous Bell Labs, set up in 1925 by the company Bell founded, and Bell Labs would spin off in turn that amazing aggregation of world-changing intellectual energy known as Silicon Valley (see page 81).

Bell would get rich from his endeavors, but his money was soon poured back into his next set of experiments; money never was the main object. As a wedding present, Bell gave Mabel 1,487 of his 1,497 shares in the new Bell Telephone Company.

THE GENIUS DETAILS

The first words transmitted over Johann Reis's "telephon" were *Das Pferd frisst keinen Gurkensalat,*" or "The horse doesn't eat cucumber salad." Indeed.

President Rutherford B. Hayes had the first phone installed in the White House, with the number "1."

There are roughly 6 billion telephone subscriptions in the world today, including 1.26 billion fixed-line subscriptions and 4.6 billion mobile subscriptions. The first mobile telephone call was made in 1946.

Bell considered his "greatest achievement" to be the "photophone," a wireless telephone he invented with his assistant Charles Sumner Tainter in 1880 that could transmit human conversations and other sounds on a beam of light—the basis of fiber-optic communications, a hundred years early.

On his death in 1922, Bell's wife, Mabel, whispered to him, "Don't leave me," and he signed back, "No," before expiring. On his death, every phone in North America went silent in his honor.

The word *decibel* was invented in honor of Alexander Graham Bell.

MAGNETIC TAPE RECORDING

M agnetic tape recording, both audio and visual, had one of the more unusual lineages of any American invention, beginning its career as a prize of war. It was made commercially viable in the United States, thanks to the contributions of two perfectly typical Americans: a former fighter pilot for the Czar of All the Russias, and one of the greatest of all radio crooners.

Alexander M. "Alexi" Poniatoff was born in 1892 near the provincial city of Kazan in the Russian Empire. His interest in mechanics led his father, a wealthy lumber merchant, to send him to Berlin for advanced degrees in engineering. Returning to Russia, he became a pilot for the Imperial Russian Navy in World War I and then for the White Russian forces in the civil war that followed. When the communists won, he fled to China to work for the Shanghai Power Company, then moved on to the United States, becoming an American citizen in 1932, at the age of forty. Working at General Electric, Pacific Gas & Electric, and the Dalmo Victor Company as an electric engineer during World War II, Poniatoff perfected a line of motors and generators for airborne radar systems.

Going into business for himself in 1944, Poniatoff founded the Ampex Corporation, named for his initials plus "excellence." Ampex set up shop in Northern California, where a sort of prototypical "Silicon Valley" sprang up after the war, centered on the defense industry and the region's superb public universities. Poniatoff succeeded in getting some top engineers to come work for him. But competition was fierce, and Poniatoff and his small team had to search constantly for new business opportunities. They found one when they heard a

Carrying the voice of Bing: the 1949 Ampex magnetic tape recorder.

demonstration of a Magnetophon, a magnetic recording device manufactured by the Telefunken corporation in Germany and literally captured in the last days of the war by Jack Mullin, a US Army major in the Signal Corps.

Mullin was an engineer in the film industry, not an Ampex employee, but he willingly shared this technology for free, believing that he had recovered it at the taxpayers' expense and had no proprietary right over it. Mullin and Poniatoff's team took the German machines apart, then built their own prototype. Magnetic recording had been theorized as far back as 1888 by the American engineer Oberlin Smith, and various forms of magnetic tape had been developed by European inventors over the decades. It had a bad reputation, though, thanks to early British tapes that were made of paper and painted with magnetic oxide. The Ampex engineers went with Germany's oxide-coated plastic film, then developed superior electrical reel and capstan motors that virtually eliminated "flutter" interference.

Their new machine, the Ampex Model 200A, was a major breakthrough. But how to get it out on the market? At the time, Ampex had only seven employees. Mullin came to the rescue again, taking the new device down to Bing Crosby's ABC studio in Hollywood. Crosby at the time was doing a live show every week on the radio—or rather *two* live shows, one for the East Coast audience and one for the West, three hours later. The crooner saw at once that with a recording device of this quality he could cut his workload in half. Crosby immediately ordered up twenty of the new recording devices at $4,000 apiece, with a 60 percent down payment. Soon Der Bingle, a shrewd businessman, was promoting Ampex recorders himself.

It was the sale Poniatoff's company needed to spring ahead, going from a seven-man concern to a global electronics giant employing thirteen thousand people in the course of the next fifteen years. The company moved rapidly in pioneering multichannel (multiple-track) tapes for use by the military, by the space program, and on movie soundtracks. It also whittled down its bulky industrial models to affordable mobile tape recorders for mass consumption.

The first video recording device, BBC's Vision Electronic Recording Apparatus, used a thin steel tape traveling at over two hundred inches per second. The amount of tape it required for even one minute of video recording made it impractical.

Ampex's successful transverse-scan technology packed much more data on every inch of tape, allowing tape speed to be reduced from two hundred inches to fifteen inches per second.

The voice of Emperor Franz Josef, of the Austro-Hungarian Empire, was recorded in 1900 at the Paris Exposition on a Poulsen Telegraphone—the oldest surviving magnetic recording.

Sony marketed the first home video recorder in 1964. Ampex and RCA followed a year later with monochrome, reel-to-reel recorders for under $1,000.

Sony's Betamax video recorder debuted in 1975—but two years later, JVC released its VHS home video recorder, which would soon come to dominate the marketplace.

Working on an eight-channel data recorder that the army could use when it blew up stuff—tanks, trucks, half-tracks—at its Aberdeen Proving Grounds, another leading Ampex engineer, Charles Ginsburg, encountered a seventeen-year-old prodigy who helped out at the company's Redwood City headquarters in between high school. Ray Milton Dolby, the son of a local inventor, soon proved invaluable in working on another preoccupation of Poniatoff's and Ginsburg's: *video* tape recording.

Their work was slowed when Dolby had to leave for his own two years of military service. The central challenge was that video signals require more bandwidth than audio signals. Mullin and the BBC had created systems that moved tape across a fixed tape head at very fast speeds, but these required enormous amounts of tape and still didn't record very well. Ginsburg and Dolby's colleague at Ampex, Walt Selsted, came up with the key innovation: four equidistant heads that would record the video signal transversely—that is, down the tape's width—as it moved at a normal speed. The heads would be at a right angle to the tape, held there by a vacuum system to ensure that tape and heads always stayed in contact with each other.

Introduced at a convention of CBS personnel in Chicago in April 1956, the video recorder brought the house down, leaving its audience "shouting, screaming, and whistling," and standing on chairs, according to one account. Video recording would revolutionize television, making prerecorded shows the norm and eventually making innovations such as slow motion and instant replays possible.

Weirdly, even though it had pioneered the consumer audio recorder, neither Ampex nor any other American company seemed very interested in producing a video recorder for home use—in a country whose citizens were already averaging seven hours of TV watching per day. European and especially Japanese firms were rapidly catching up in magnetic recording technology, and they would come to dominate, then monopolize, the VCR industry in the 1980s and '90s. It was an opportunity missed by a US industry that had lost the vision and drive it possessed back when it was still just a handful of engineers in a room, trying to make Bing Crosby come through in all his glory.

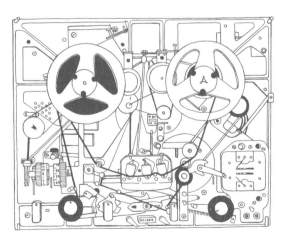

The guts of a three-head 1964 Ampex Fine Line F-44 audio tape recorder for home use. Ampex engineers kept the oxide-coated plastic film German engineers pioneered, then added superior electrical reel and capstan motors to all but eliminate "flutter" interference.

"A COMPUTER ON A CHIP"
THE MICROPROCESSOR

A mericans alive just after World War II thought of computers as "the Great Brain," the US Army's Electronic Numerical Integrator and Computer (ENIAC)—a thirty-ton, 2,400-square-foot behemoth relying on no fewer than 17,468 vacuum tubes to perform its calculations. The story of how computers got small—and therefore how they could become ubiquitous compact instruments of almost limitless power, run by countless microscopic transistors—would be the technological story of our age. It would also exemplify the singular American ability to create dazzling nexuses of invention and application and to draw the best minds from around the world to such efforts.

The transistor itself was invented at one such nexus, the invention mill known as Bell Labs in Murray Hill, New Jersey, spiritual descendant of the great Alexander Graham Bell's Volta Bureau (see page 74). The Nobel Prize–winning work was directed by William Shockley, the acerbic genius who proved in 1947 that the bulky, easily overheated vacuum tubes that computers then relied on could be replaced by three semiconductors layered together. ENIAC's vacuum tubes could switch circuits on and off ten thousand times a second. Transistors could do this *billions* of times a second—and were just one-fiftieth the size!

A computer on every desk: what the microprocessor made possible.

The trouble was that Bell Labs' parent company, AT&T, already enjoyed a monopoly on the nation's phone service, and the US Justice Department was not about to let it have one on computers, too. Bell Labs trained and licensed twenty-five domestic and ten foreign companies in the use of transistors for the relative pittance

THE GENIUS DETAILS

Gordon Moore, Noyce's longtime partner, is credited with "Moore's Law" for his prediction in 1964 that the number of transistors on a microchip would double every year.

At the height of the space race, NASA was buying up 60 percent of all the microchips made in America. The Apollo Project (see page 37) bought over one hundred thousand of them in 1964 alone.

Fairchild engineers were able to fit 1,024 transistors on a single integrated circuit by 1967. Soon after, they reached 1,024 bits—or 1 K of "random-access memory" (RAM), giving them nearly the entire memory chip market.

The first 4004 microprocessor chip was made up of 2,300 transistors. By 2015 the highest number of transistors in a commercially available chip was 5.5 billion.

In October 1971 Intel's initial public offering raised $6.8 million, or $23.50 a share. Playboy Enterprises, then at the height of its popularity, went public on the same day at the same price. Within a year, though, Intel's stock price was twice that of Hugh Hefner's company, leading a Wall Street analyst to say, "It's memories over mammaries."

of $25,000 a pop (free to anyone developing transistors for use in hearing aids, a tribute to Alexander Graham Bell's long quest to help the deaf to hear).

Shockley left Bell to set up his own company where he'd grown up, near Palo Alto, California. He brought a sizable contingent of Bell's talent with him.

Nobody knew it yet, but this was the start of Silicon Valley. It was perfect: great weather and proximity to some of the world's best universities. The only fly in the ointment was Shockley, who proved to be a hectoring, imperious, and occasionally unbalanced boss. Eight of his scientists—whom he labeled "the traitorous eight"—soon departed, under the leadership of twenty-nine-year-old Robert Noyce.

Noyce hailed from what had been an earlier American nexus of enlightenment, Grinnell College in Grinnell, Iowa, where before the Civil War eleven Congregational ministers from Massachusetts had set out to battle for the Union and the Lord and against frontier slavery. The son of a minister himself, as well as a champion swimmer and all-around big man on campus, Noyce discovered his life's work when his physics professor showed him two of the first transistors ever made.

A natural leader, Noyce found an investor for "the traitorous eight" in Sherman Fairchild, an inventor and founder of a camera company. Fairchild Semiconductors would provide the foundation myth of Silicon Valley, setting the template for electronics geniuses. Other than reporting to work by eight, there were no rules and no rigid hierarchies or executive privileges. The company would spawn over one hundred "Fairchildren" start-ups in less than twenty years, while Noyce became known as "the Mayor of Silicon Valley."

This happy band of anarchists was directed, ironically, by the needs of America's defense and aerospace industries, which bought up much of their product. It became "Silicon Valley" because silicon was the only transistor coating able to withstand the heat of nuclear missiles as they fell back into the atmosphere. The need to pack computers onto warplanes and spaceships provided a constant impetus toward making transistors ever smaller and lighter—which in turn made them more and more difficult to effectively wire together.

This challenge was answered in 1958–59 by Noyce and a towering, good-natured young scientist named Jack Kilby, who was toiling down in Dallas at Texas Instruments. The answer was to put *all* the transistors together in a single, integrated, "monolithic" circuit. Noyce eliminated connecting wires altogether, printing lines of metal on the transistors. But the chip wasn't done yet. Noyce and his longtime partner, chemist Gordon Moore, were now in another feisty start-up they called "Intel" (mostly because "Moore Noyce" did not work for anything this side of a heavy-metal group). One of their first hires, Marcian "Ted" Hoff, an electrical engineering prodigy from Rochester, New York, hit upon the next evolutionary step: a single chip capable of "mixing," as T. S. Eliot put it, "memory and desire"—putting all commands *and* memory storage on a single circuit.

The microprocessor—what Intel would proudly dub "a computer on a chip"—was born. Its first model, the 4004, was a primitive technology by current standards but—at just one-eighth of an inch wide and one-sixth of an inch long—already as powerful as ENIAC. A more polished successor, the Intel 8080, would empower the first personal computers.

This is, by necessity, a greatly condensed version of how the microprocessor was invented and by whom. Silicon Valley, even sixty years ago, was already attracting genius from around the world. Half the original "traitorous eight" had fled from oppression or lack of opportunity in Europe. Key contributors to the microprocessor included a chemical engineer from Japan, Masatoshi Shima, and the twenty-eight-year-old physicist son of a Virgil scholar from Italy, Federico Faggin. The microprocessor's roots lie deep in the best of American character: ambition, vision, and smart laws spurring competition—not to mention men of faith battling slavery out in the prairie town of Grinnell; and an earlier immigrant genius inventor, Alexander Graham Bell himself, whose greatest hope was that he might allow his wife to hear.

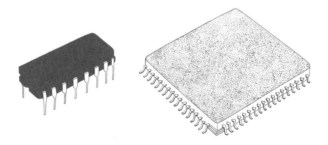

At left is Intel 4004, the first microprocessor; to the right, a fourth-generation microchip of today, capable of containing 5.5 billion transistors in less than half an inch of space.

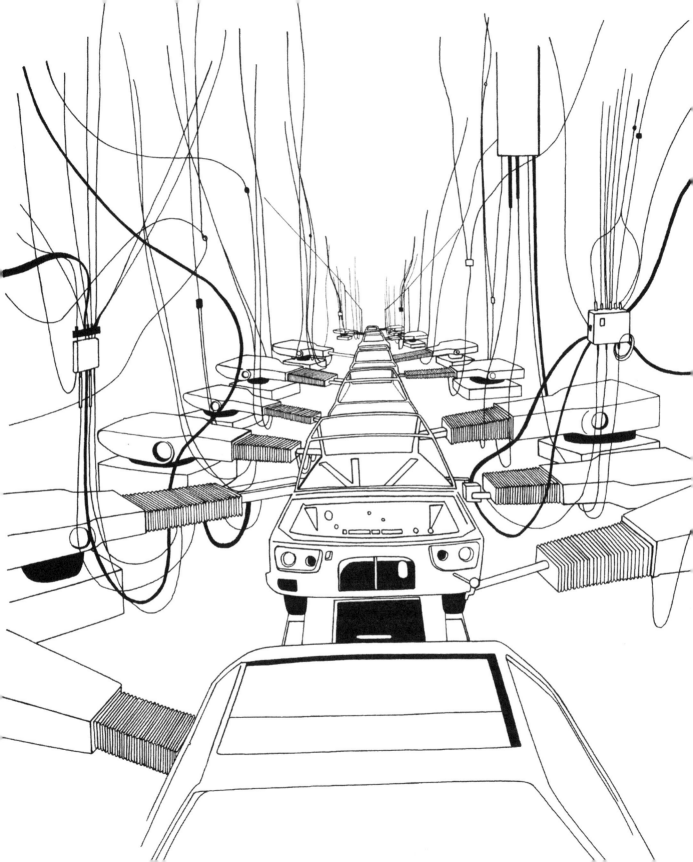

THE FIRST INDUSTRIAL ROBOT

A t New York's 1939–40 "World of Tomorrow" World's Fair, people lined up for hours to see "Elektro," a seven-foot-high, 265-pound robot, give a twenty-minute demonstration of the future. Elektro, who was eventually joined by his mechanical dog, Sparko, could walk; move his arms, head, mouth, and eyes; talk; and smoke a cigarette. The more cynical observers felt sure there was a man hidden inside Elektro's bronze-colored, brushed aluminum skin, but his creators at Westinghouse had made sure to put a hole right in the middle of Elektro's stomach to show there was no such thing.

Elektro was a considerable technological accomplishment for the time, motivated by a combination of camshafts, gears, and motors. An outside operator could convey signals to the robot through the vacuum tubes, photoelectric cells, and telephone relays inside him. A bellows system let him smoke the lit cigarette, and a .78 record provided his vocabulary of seven hundred words.

Audiences were enthralled. But they were missing out on the real action: the automated photoelectric counters at the World's Fair's entrance, there to count every paying customer entering the future—that is, the "electric eyes" that now routinely open so many shop doors for us. Invented by George Charles Devol Jr., they were one more advance toward developing the world's first real industrial robot.

Devol was born to a family of means in Louisville, Kentucky, in 1912. He went to prep school, but like so many inventors his tastes ran less to books and more to how things worked. He loved tinkering with the electrical and mechanical systems of airplanes, boats, engines, and his school's new electrical plant. Above all, he was interested in seeing what else vacuum tubes—those fragile, bulky, glass tubes that controlled the passage of electricity between electrodes—could do, besides power radios. Devol decided to skip college and go right into business, starting a company he called United Cinephone, to better coordinate film and sound in the new "talkies" that were taking over the nation's movie palaces (see page 239). Like many inventors,

The Unimate robot is seen here on the General Motors assembly line in Lordstown, Ohio, where in 1969 it helped produce 110 cars an hour—over twice the rate of any plant then in existence.

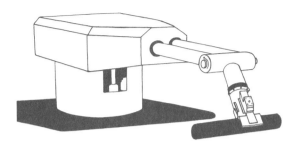

Unimate lifting a door handle.

he had taken an early wrong turn. The electrical giants RCA and Western Electric already had that field monopolized.

Casting around for a new mission, Devol began designing photoelectric switches and licensing them to Yale & Towne, which in turn invented the first doors that opened automatically when a person approached (something the company marketed under the romantic name of the "Phantom Doorman"). He went on to use photoelectric cells and vacuum tube controls to create a bar code system nearly half a century before such systems became ubiquitous.

World War II put this work on hold, but after the war Devol developed digital magnetic recording for business data (see page 78) and high-speed printing systems. His magnetic recording system was too slow for business, so he reused it as a machine control—the brains of the first robot.

Robots were supposed to look like Elektro, humanoid shapes of metal that walked and talked. What Devol had invented instead, by 1954, was the Programmed Article Transfer, a mechanical arm attached to two boxes, one of which held its memory. This was the concept of "Universal Automation"—or "Unimate," as his wife, Evelyn, convinced him to call it.

It was something new in the world—a machine that could repeatedly perform tasks from memory and be reprogrammed to perform new tasks. Devol's patent application required not a single prior citation.

And no one would take a chance on it. The problem seemed to be the sales patter worked up by Devol, who could not keep himself from spouting out every thing he could conceive of robots doing in the future. If science fiction mavens wanted Elektro, business magnates already considered robotics pie in the sky.

The only one who would listen was thirty-one-year-old Joseph F. Engelberger—a Columbia PhD in engineering and a science fiction devotee. He was chief of engineering in the aircrafts products division of Manning, Maxwell and Moore (MMM), a manufacturing firm in Stratford, Connecticut. He was not put off by Devol's sales pitch—but no sooner had he heard it than MMM was bought out and Engelberger was offered a promotion if he would dissolve his division.

Engelberger turned the offer down and instead found a new backer for Devol's idea at Consolidated Diesel Electronic (Condec). Then he and Devol set to building the first Unimate robot. Its very novelty made the task all the harder. They switched from bulky vacuum tubes to those brand-new magic chips of conductors known as transistors (see page 81) for controls but still found that other parts, such as digital encoders (tiny discs, coded with photographically deposited radial patterns, that convert motion into sequences of digital pulses) were simply not up to what Unimate was supposed to do. They had to build Unimate's hydraulic parts themselves. Engelberger recruited a team of crack engineers to design and machine whole new technologies, such as a rotating drum memory system, under Devol's direction.

After $5 million and seven years of hard work, Devol personally sold the first Unimate robot, which went on the line at General Motors' Inland Fisher Guide Plant in Ewing Township, New Jersey. There were—and would continue to be—many fears that such robots would quickly replace men. But Unimate's first job was handling die casting—moving molten pieces of car metal straight from the die-casting machine and stacking them on the assembly line, work that was dangerous and difficult and that autoworkers hated to do.

Over the years, its inventors would continually improve on and add to Unimate, producing robot arms with six fully programmable axes of motion, capable of handling parts weighing up to five hundred pounds. By 1966 Unimate was in full production in Connecticut, and by 1975 it showed its first profit as it made robots for all auto companies to handle and weld car parts. Automation was here. It would indeed eventually replace many of the manual jobs on the assembly line, and Unimation would decline and be replaced itself in the 1980s as robots moved from being hydraulic to electrical. But cars would now be manufactured more quickly, precisely, cheaply, and safely than they ever had been before.

THE GENIUS DETAILS

The word *robot* was invented in 1921 by Karel Čapek, who used it in his play *R.U.R.* ("Rossum's Universal Robots").

Popular science and science fiction writer Isaac Asimov coined the word *robotics* in a 1941 short story, "Liar!" that appeared in *Astounding Science Fiction* magazine. The first Unimate robot was named "Isaac" in his honor.

George Devol's early experiments with microwave oven technology led to his creation, shortly after World War II, of the "Speedy Weeny," a microwave machine that cooked and dispensed hot dogs on demand at locations such as Grand Central Terminal.

Unimate made a big splash when it appeared on *The Tonight Show* in 1966 and delighted host Johnny Carson by putting a golf ball into a cup, opening and pouring a beer, waving the bandleader's baton, and grabbing an accordion and waving it around.

3-D PRINTING

I t seems, on the surface, to be a childishly simplistic idea. An alchemist's dream, akin to Rumpelstiltskin spinning straw into gold, or the transformers on *Star Trek*: the idea that you can simply "print out" a wholly usable, three-dimensional object. No, 3-D printing does not actually change the molecular structure of a substance—but it *does* alter its particulate nature, almost always from a liquid to a solid.

A form of primitive 3-D printing goes back as far as nineteenth-century photo-sculpture, when French artists would photograph a subject from every angle, then project those photos onto some solid material, enabling them to carve a sculpture from it. Starting in the 1950s, a number of US companies and individuals began experimenting with ways to use light to shape photopolymers into three-dimensional images and even objects. They had little success.

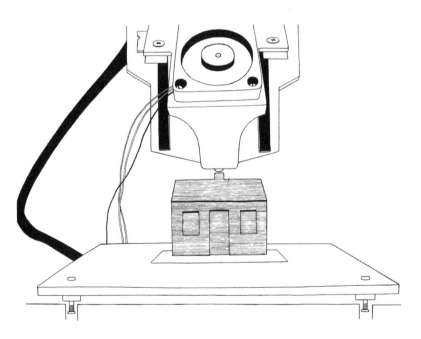

They are making more real estate: an instant house being "printed" out.

But early in 1983, Chuck Hull, a genial design engineer from Colorado who was trying to use ultraviolet lamps to put plastic veneers on furniture, grew frustrated by how long it took to get parts for new product design prototypes and decided to try to "print" them himself. After months of experimentation with what CNN would describe as "plastic-y gloop" in a backroom lab after working hours, Hull was able to produce a small, black eyewash cup.

Chuck phoned his wife and told her to come down to the lab and see what he had created. Already in bed in her pajamas, she told him, "This had better be good!"

It was. Hull had invented stereolithography (SLA). The way it works is that a software program known as computer-aided drafting (CAD) scans an existing product, or designs a new one, in 3-D. The design is then "sliced" into hundreds or thousands of horizontal layers. Ultraviolet lasers trace this design onto a vat of photopolymer resin, layer by layer, curing and hardening the layers as they move down the vat.

Over the last three decades, all sorts of related processes have sprung into being, utilizing materials ranging from sheet metal to ceramics to pasta and chocolate. The machines necessary to handle these materials vary, of course, in their size and power, but the idea is the same. Already, a battery has been printed at Harvard, an electric car in Chicago, houses in China and Amsterdam. In the very near future, 3-D printers will recycle your plastic milk containers; print out your medicines, and maybe your food; create your clothing, eyeglasses, and jewelry.

It is a technology that combines nearly all the roles of a huge manufacturing enterprise. Instead of having separate departments design a new car, sculpt the dies (molds) for its parts, fashion the necessary materials, and then put everything together, you can have a computer and its "printer" do it all for you (some assembly may still be required). Another way to understand it is by the alternative name for 3-D printing: additive manufacturing. That is, instead of traditional, *subtractive* manufacturing—in which, say, an object is carved out of a big hunk of metal—everything is *added*, often from scratch and almost instantaneously.

The applications are almost limitless. 3-D printing is at a premium in anything requiring detailed, customized work and rapid innovation: prosthetics and implants that fit better than ever; rapid prototypes of new running shoes—or spaceship parts, or jet engines; entire human organs that can be made from living cells, squeezed layer by layer out of a nozzle; maybe, someday, a habitat on the moon.

3-D printing will change *how* we do business. As the price of industrial printers continues to drop from $20,000 to below $1,000—and the cost of smaller, personal printers drops below $100—we will witness the continuing rise of "rapid manufacturing," "on-demand manufacturing," even "desktop manufacturing." There may soon be a 3-D printer app available—if it's not already.

François Willème, a Parisian artist, invented photosculpture in 1859 by having a subject photographed from twenty-four different angles by twenty-four different cameras, projecting the images on a screen, then using a pantograph to cut a 3-D wood sculpture of the images.

The Strati, the first completely 3-D-printed car, was printed in the space of forty-four hours by the Phoenix automaker Local Motors. Made out of fiber-reinforced thermoplastic and powered by electricity, it weighs just 450 pounds, with only 40 components, and claims a range of 100 to 120 miles and gas mileage of 40 miles to the gallon.

Scientists first bio-printed organs from a patient's cells in 1999, a blood vessel in 2009, and a prosthetic jaw in 2012. Chinese scientists now bio-print human ears, livers, and kidneys with living tissue. Scientists in Bahrain print sandstonelike structures to help restore damaged coral reefs.

"Three-dimensional printing makes it as cheap to create single items as it is to produce thousands and thus undermines economies of scale. It may have as profound an impact on the world as the coming of the factory did," the *Economist* has pointed out.

The implications of this are enormous. Some have gone so far as to call it the backbone of the "Third Industrial Revolution." Production lead times could be slashed by more than half. Additive, onsite manufacturing could save enormous amounts of waste and pollution by building goods efficiently and obviating the need to transport them over long distances. It could reverse the decades-long migration of manufacturing to low-wage countries—though it may also eliminate untold numbers of jobs everywhere. Worldwide, the 3-D printing industry is expected to increase sevenfold, from $3.07 billion in revenues in 2013 to over $21 billion by 2020.

Chuck Hull, now the holder of ninety-three patents and a wealthy man, thanks to his 3D Systems company, says he is "humbled" by what his invention has brought about, particularly in the field of medical technology. He originally predicted that it would be "25 to 30 years" before 3-D printing took off, but as early as 1996 surgeons used his invention to make a model that helped them separate conjoined twins—something he described as "really touching for me." He continues to work in the field well into his seventies because he finds it "so interesting. It's a really interesting journey." And it's just begun.

THE EVOLVED HUMAN
THE CYBORG

Modern people in the first modern nation, we are always changing, even physically. Brain implants to treat epilepsy and Parkinson's disease with "deep brain stimulation"—a sort of brain pacemaker—and cochlear implants for the deaf and hard of hearing have been in use for almost twenty years, and retinas are now being implanted. The next *big* thing to transform human existence may be just around the corner: our "enhancement" by computer-connected microchips.

Self-proclaimed cyborg and artist Neil Harbisson, a European transplant to New York who was born completely color-blind, already uses an antenna mounted on his head and attached to a microchip in his skull; the electronic waves convert colors into sounds that he hears via brain conduction. His partner, a performance artist who calls herself Moon Ribas, has a microchip in her arm that supposedly can sense earthquakes anywhere in the world.

Amal Graafstra, a self-styled "adventure technologist" from the state of Washington, had radio-frequency identification (RFID) tags—or minitransmitters—installed in the web of each hand over ten years ago, enabling him to open his garage and front doors, turn on his computer, and start his car and motorcycle—at least if he gets really, really close to them (his chips' range is only two inches).

If these seem more like gimmicks than anything else, well, they mostly are. But just as early mechanical television (see page 243) or phonographs (see page 218) preceded inventions that became ubiquitous (and all-consuming), the enhanced human being is likely to become a reality before the twenty-first century is over.

What scientists are looking at now are electrodes installed on the hippocampus—the part of the cerebral cortex where short-term memories are converted into long-term ones—and used to help victims of stroke, or localized brain ailments, to retain vital parts of their personalities. Already, brain implants can help quadriplegics manipulate keyboards and robot limbs. Restoring—or adding—sight, hearing, mobility, and speech to the victims of terrible accidents and maladies, they may well also control or eliminate depression, insomnia, and alcoholism. Brain implants will not be the only enhancements. Someday soon, we may have artificial devices installed all over our bodies, monitoring and improving every aspect of our health.

Government, military, and corporate entities in the United States are now at work developing all these potential additions to our infrastructure. Stanford University's "Neurogrid," developed by Ghanese American professor Kwabena Boahen, is a primitive "artificial brain" that uses sixteen linked-together IBM chips to generate one million neurons and billions of neural connections, or "synapses." IBM's new "True North" computer chip, which is the size of a postage stamp and equipped with 5.4 billion transistors, can simulate 1 million neurons and 256 million synapses.

"Brain implants today," psychologist Gary Marcus and neuroscientist Christof Koch claimed in the *Wall Street Journal* in 2014, "are where laser eye surgery was several decades ago. They are not risk-free and make sense only for a narrowly defined set of patients—but they are a sign of things to come."

The risks enter whenever someone drills a hole in your skull, however skillfully. Infections and neural bleeding are just two of them. Implants must be small, nontoxic, and energy efficient—or they will literally fry your brain. (They will, perhaps, be recharged by an "induction coil" nightcap.) It may soon be possible to sprinkle "neural dust"—thousands of microsensors, about the thickness of a hair—on the cerebral cortex to communicate, through ultrasound, with outside computers for the whole of a human life.

More daunting is the natural technology of the brain itself. True North can simulate 1 million neurons and 256 million synapses. The human brain has some 100 million neurons and maybe up to 1 *quadrillion* synapses—all of which it runs at just 1/40,000 of the power it takes to keep a personal computer humming.

Currently, neurologists can read the regions of the brain where activity takes place, but we still know little about how exactly individual neurons interact. This is the object of the new science of optogenetics, which proposes to manipulate the brain at a molecular level, and which is already being used to make retinal implants. In optogenetics, brief pulses of colored light use the genes in every neuron to control the mind—"effectively turning the brain into a piano that can be played," as Drs. Marcus and Koch put it.

Do we *want* our brains played like fine instruments?

Ethical questions abound. Most of us are fine with using implants to help the miserably burdened, but there's a sliding line. Will we get implants to help the blind to see and the deaf to hear—or to be able to see in the dark and hear through our neighbors' walls? Will we just end depression or make ourselves feel good all the time? Can we really trust today's competitive parents not to fill their children with implants designed to make them concentrate better on tests or throw a perfect spiral on the football field? And if only the rich and the well-connected can afford such advantages, will we become truly different species?

"Biohackers" and "grinders" are thrilled by the idea of being hooked up permanently to the Internet, able to summon up whole libraries of information before their mind's eye with a simple, mental command. The new miracle chips might let you make a phone call in your brain, send a text, summon a work file, even improve your golf game. And further down the road is "the rapture of the geeks," our wonderful new brains preserved in some sort of electronic casing and carried on forever in newer and better bodies.

Potential dystopias are more easily imaginable, even if you don't believe all the chips will bear the number 666. They seem the very model for a modern dictatorship. And if we'll all be connected, all literally on the grid, how easy would it be to take us all down?

Apart from all the disaster scenarios, will you really enjoy having a book "inserted" directly into your memory, instead of having the experience of reading it? Chips talking to chips promises to create a sort of telepathy, which could be very useful, and also maddening. Do you want to walk down a street and hear what everyone is thinking—or *not* thinking—about you? (And talk about drunk dialing!)

Intel believes that your brain will be connected to your computer by 2020 and is currently hard at work researching ways to read brain waves so that they can operate computers, TVs, and cell phones—through Intel sensors.

"When you think about something and don't really know much about it, you will automatically get information," Google CEO Larry Page has been quoted as saying. "Eventually you'll have an implant, where if you think about a fact, it will just tell you the answer."

Yes, but whose answer?

THE GENIUS DETAILS

Today over one hundred thousand people worldwide already have brain implants, all of them for medical reasons. Some eighty thousand of these are used to treat epilepsy and Parkinson's disease. Over three hundred thousand people worldwide have cochlear implants.

"Enhanced humans" may be able to use a "sixth sense," a magnetic navigational system. Minnesota engineer Brian McEvoy had a subdermal compass implanted in his brain in 2013.

The BrainGate system allows victims of paralysis to gain some control of outside functions by installing a small, brushlike chip, with some hundred tiny wires, through the skull and into the motorcortex, which controls movement.

DARPA (Defense Advanced Research Projects Agency) is working on a wearable "exoskeleton," thought-controlled robots, "thought helmets" to make telepathic communication between soldiers possible, and brain-computer interfaces (BCIs) that will enable troops to see in the dark and to detect land mines.

OUCH!
THE SAFETY PIN

I f Walter Hunt was not the most prolific inventor in the first decades of the United States, he was surely the most maddening. No doubt, the list of what he dreamed up has left generations of would-be magnates foaming at the mouth over what *they* would have done with Hunt's many inventions, any one of which might have made him a fortune.

Hunt has been described as "a Yankee mechanical genius," though perhaps a mechanical Forrest Gump would be closer to the mark. Good-natured, philanthropic, and always caring, he was one of thirteen children born to Quaker parents in 1796, in the little upstate New York town of Martinsburg. Taught in a one-room schoolhouse, he is reported to have earned "a degree in masonry," although where he would have done that or what it would have meant in the early nineteenth century remains mysterious.

Hunt might have been content to spend his whole life outside of Lowville, another small Lewis County town, farming to support his wife and four children, but the local flax mill—where a brother worked—got into trouble. Hunt responded by inventing a better spinning and roping machine that he let mill owner Willis Hoskins and his associate, Ziba Knox, take out the patent on. Well, how can a man deny anything to someone named Ziba Knox? Not long after, in 1826, Hoskins was threatening to slash all his workers' wages because his mill was floundering again. Hunt talked him down and invented a still better flax spinner for him, although this time, in an inexplicable burst of self-interest, he patented it for himself and set off with his family to try his luck in New York City.

On his first day in the big city, Hunt watched a young girl run down in the street by Manhattan traffic that was even more murderous than it is now. Shaken, he promptly invented and patented a metal gong alarm that coach drivers could operate with their feet, rather than having to take their hands off the reins to blow a horn. He then sold the device to a manufacturer for a pittance.

So it would go, Hunt inventing one incredibly useful and commercial device after another in the little machine shop he kept in an alley off Abingdon Square. A knife sharpener, with a safety guard—Hunt, unlike everyone else in nineteenth-century America, was all about safety first—that could replace the huge, traditional grindstones that were too big for apartment living. A rope-making machine. Castor globes (those little luggage wheels) for moving furniture. A coal stove that distributed its heat equally in all directions. And more: an iceboat, a saw, a conical bullet, a revolver, a breech-loading, repeating rifle (see page 185); several bicycles, a paper shirt collar, a nail-making machine, a flexible spring attachment for belts and suspenders, hobnails for boots and shoes, bottle stoppers, a safety lamp, a "ceiling-walking device" for the circus, an ink-stand, and the modern fountain pen. Some twenty-eight patents in his lifetime, each one them capable of making an antebellum American leap up and say, "Hell, yeah! Why didn't somebody think of that already?"

Yet what Hunt invented he didn't always patent, what he patented he usually sold for a small flat fee to pay some pressing debt, and what he hung on to he rarely bothered to exploit. The most notorious case was the flawed but novel sewing machine (see page 97) that he invented. First he sold the rights to a manufacturer, who could not raise the money to market it. Then he decided to shelve the whole project because his fifteen-year-old daughter, Caroline, convinced him that it would put seamstresses out of work. (The truth would be exactly the opposite.)

Even as his family fell in and out of debt, Hunt seemed to prefer working on rather nauseatingly precious paintings of puppies, ponies, and other farm animals and speculating in land along the Hudson River to doing anything with his inventions. (Perhaps the greatest mystery of Walter Hunt is why he thought he had the business head for being a real estate mogul.)

By 1848 Hunt found himself in debt again, owing fifteen dollars to a draftsman named Jonathan Chapin. Hunt being Hunt, *he* went to see Chapin, who, far from pressing him for the cash, offered to loan him more. Hunt, in despair, refused—then looked at the length of brass wire he had been twisting about in his agony. Eureka! Hunt hustled back to his machine shop, bent the pin around, and invented the safety pin.

Paleontologists would estimate that some form of safety pin or another had been around since the Neolithic Period. Yet somehow man had never got it right—until Walter Hunt. His was a masterpiece of simplicity. He shaped the very first safety pin with a completely shielded point and a spring action to tuck it easily inside the guard. He then hurried out to a manufacturer named Jonathan Richardson and offered to sell it to him for $400. Richardson

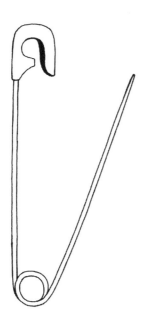

Walter Hunt's greatest invention. A single factory today can churn out a billion safety pins in one year.

THE GENIUS DETAILS

The ancient Greeks and Romans used a sort of safety pin, called a "fibula," or brooch. Etruscans and Persians also used some variety of this. But no previous pin had Hunt's clasp-and-spring action.

Hunt did not come up with the name *safety pin* but filed his invention patent as "a new and useful Improvement in the Make or Form of Dress-Pins."

Made mostly out of steel in the nineteenth century, safety pins today are made out of cheaper brass, sometimes with a chrome coating.

In India, safety pins and sewing needles are often passed down over generations.

Hunt's device for helping circus performers to walk upside down would still be in use in 1937, roughly a hundred years later.

agreed—*if* Hunt went to the time and expense of patenting it and gave him all rights. Ecstatic, Hunt said yes, and rushed back to Chapin to give him his fifteen dollars. Fortunately, he did not encounter anyone offering to sell him magic beans along the way.

Hunt would expire of pneumonia in his Abingdon Square workshop in 1859 at age sixty-three. Before he went, he at least had the satisfaction of knowing that Isaac Singer, wrapping up all loose ends in the epic sewing machine patent battle, had agreed to pay him $50,000. In 1875, the Union Paper Collar Company also agreed to pay his son, George W. Hunt, $5,000 plus court expenses in cash, $50,000 in company stock, and 10 percent of all royalties for the paper collar Walter had invented.

Or maybe, for Walter Hunt, who was never heard to complain or bemoan his luck, the satisfaction came in the work. Babies all over the world, at least, would be eternally grateful.

THE SEWING MACHINE WARS

People have been sewing ever since the last Ice Age, twenty thousand years ago, when they used bone needles to bind skins and furs together. Over all the millennia that followed, no one came up with a significantly better process; even Leonardo Da Vinci tried and failed.

Beginning in 1755, a number of European inventors began devising one form of sewing machine or another, but they never caught on. This was for many reasons, but the main failing was that none of the machines succeeded in making more than a simple, weak "chain stitch" that would easily tear out.

This changed in 1834, when Walter Hunt invented a device that contained all the basic elements of the modern sewing machine in his Manhattan shop. Hunt, a sort of business idiot savant (see page 185), refused to even patent his sewing machine because his fifteen-year-old daughter was afraid it would put seamstresses out of work.

A few years later, Elias Howe took up the quest. He came from a family with a long history of invention but was put to work on the family farm at six and was rented out to work a neighbor's fields by eleven. Married by twenty-two, with children soon on the way, he worked full-time for nine dollars a week while spending his spare time trying to make a sewing machine. A school chum backed him with tools and a small annex to his

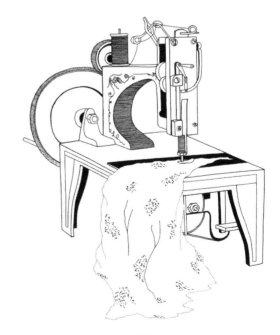

The classic design that was the 1854 Singer sewing machine.

house to use as a workroom, and after four years and forty prototypes Howe completed a patent model.

The eye of Howe's machine needle was at the point of the needle, not the top. As it plunged forward, into the fabric, another thread, held on a metal shuttle, passed through the loop in the first thread. The needle then retracted, pulling the threads together into a sturdy "lock stitch."

Howe had succeeded where Da Vinci had failed. Once he had his patent, he tried to publicize his new machine, holding a public contest at Boston's Quincy Market in which he competed against five women seamstresses. Howe's machine completed 250 to 300 stitches a minute without breaking a thread, while the women could hand-sew no more than 40. Still, there were no takers. His invention cost a hefty $300, and it had a balky feed—probably thanks to the fact that the material was fed in vertically and held by a delicate set of pins, while the needles worked horizontally.

Howe took his invention to England for the next three years, where he was bamboozled like the Yankee rube he was. Receiving word that his wife was dying, he sold the latest version of his sewing machine and his papers to get passage back to New York. He made it to his wife's deathbed just in time—only to get word that another ship, carrying the last of his possessions, had sunk off Cape Cod.

Yet unbeknownst to Howe, while he was in England, other American inventors had done much of his work for him, dramatically improving his machine. Chief among them was one Isaac Merritt Singer, as great a cad as the ranks of American inventors have ever produced. The son of poor German immigrants from upstate New York, Singer ran away from the family farm at age eleven to join a traveling stage act. He would perform as an actor for years, off and on the stage, a real-life Falstaff.

Yet in his peripatetic existence Singer had acquired considerable mechanical ability and business acumen. Invited to work on another inventor's sewing machine, he made critical improvements to it over a few months of feverish round-the-clock work. Most crucially, he mounted the machine on a table and reversed its direction: the fabric would be fed in *horizontally*, held by a metal plate, and operated by a foot treadle instead of the hand crank that Howe and all previous sewing machine inventors had used. The needle would be held by a fixed arm and would now move vertically, striking *down*, while a falling shuttle, beneath the plate, would feed the "locking" thread into the stitch.

Maneuvering in his usual fashion, Singer cheated and bullied his two backers into surrendering their rights, then used all his showmanship to market the device he now owned—often singing the mournful ode to the sewing girl's life "The Song of the Shirt" while exhibiting his work. Rival companies and inventors kept popping up everywhere, but Singer seemed to enjoy fighting them off.

Then Elias Howe showed up, demanding Singer pay him $25,000 for the rights to his patent. It was cheap at a hundred times the price, but Singer responded by threatening to kick Howe down the steps of his shop. He had misjudged his man. Broke and struggling to raise his three young, motherless children, Howe took Singer to court, waging a five-year legal battle that went all the way to the Supreme Court—where he won.

The judgment seemed to sober Singer, who by now had a partner: a cold, calculating lawyer named Edward Clark who had helped him cheat his old collaborators. The sewing machine business was already sliding toward anarchy, with the leading companies engaged in more suing than sewing—what the newspapers labeled "the Sewing Machine Wars."

A grand compromise was reached: the seven leading manufacturers would pool all patents and put fifteen dollars for each machine sold into the "Great Sewing Machine Combination." Howe would get five dollars on every sale—something that made him a very rich man.

It was a cartel, but one in which the members went on competing. Singer kept turning out machines that were lighter and cheaper, able to sew every possible kind of stitch in every possible kind of garment. When scandals in his personal life (mistresses, multiple wives, illegitimate children, wife- and child-beating, for starters) chased him to Europe, Clark's savvy business innovations kept the Singers selling.

The sewing machine would become ubiquitous in households, cutting the time women put into sewing for their families by nine-tenths. Singers—handsome, black, embossed machines in their felt-lined cases—could be found everywhere in homes into the 1970s.

Contrary to the fears of Hunt's daughter, the sewing machine created countless new jobs. Ready-to-wear garments democratized clothing in the United States, leaving European visitors marveling at how (relatively) well the working and middle classes dressed. They enabled millions of immigrant women to break out of the household by the turn of the century, carting their light, affordable machines around to factories and sweatshops. There they worked under horrible conditions, making what were barely survival wages. But in response these now independent women would form one of America's most powerful unions, transforming their own lives and the nation they lived in.

THE GENIUS DETAILS

According to *Godey's Lady's Book*, it would take a skilled seamstress (or housewife) about fourteen hours to make a man's dress shirt and ten hours to sew a simple dress by hand. With the sewing machine, these times would drop to one and a quarter hours and one hour, respectively.

Howe's original price for his sewing machine, $300, would be about $14,000 in today's money.

The Singer Sewing Company introduced the first electric sewing machines in 1889.

The International Ladies' Garment Workers' Union (ILGWU) was formed in 1900, with thirty dollars and two thousand members, most of them women.

MADAME DEMOREST'S FASHION SENSE

Long before the devil wore Prada, long before Donna Karan or Vera Wang or Diane von Furstenberg, Madame Demorest was telling American women how to dress and bringing the fashions of the world to their sewing rooms. At the same time, she was busy trying to abolish slavery, win equal rights for women and people of color, and stamp out drinking.

Ellen Louise Curtis was born in the New York village of Schuylerville in 1824. When she turned eighteen, her father, a hat factory owner, helped her set up a millinery shop in what was then the spa resort of Saratoga Springs. The shop was a success, and the intrepid Miss Curtis moved her business down to the much bigger town of New York City. There she met a widowed dry goods merchant, William Jennings Demorest, who had just opened "Madame Demorest's Emporium of Fashion" on lower Broadway. Ellen soon took up the role of Madame Demorest, both in the shop and in reality.

From the beginning, "Nell," as she was known, had that rare sense for fashion that, through the years, only the top designers and arbiters of style have possessed. She knew both what women wanted and what they should want, and she knew how to get it to them. Her designs included comfortable corsets—in an age of whalebone torture (see page 107)—small hoop skirts, and, most ingeniously, "Mme. Demorest's imperial dress elevators," a set of weighted loop fasteners that enabled women to discreetly lift the corners of a full hoop skirt while passing over the prodigious piles of muck lying in wait off Manhattan curbs.

The invention that changed her life, though, sprang to mind one day when she watched her maid cutting dress patterns from brown wrapping paper. *Why not have dress patterns printed on cheap tissue paper, which could then be sold to women all over the country at little cost?* She quickly made this a reality—offering the patterns along with her "Excelsior" drafting system, a mathematical formula she invented with the help

The full hoop skirt of the 1850s conceals Madame Demorest's "imperial dress elevator," an array of weighted strings and loops inside that allowed women to subtly raise the hem of their dresses while crossing the street.

Madame Demorest's first known tissue-paper pattern, for a boy's jacket, was stapled inside the pages of *Frank Leslie's Ladies' Gazette*. Demorest advertised in this publication and others, such as the popular *Godey's Lady's Book*.

The Demorests' own monthly magazine, *Demorest's Illustrated Monthly and Madame Demorest's Mirror of Fashions*, eventually reached a circulation of one hundred thousand.

Ebenezer and Ellen Butterick began producing paper patterns in a wide variety of sizes in 1863, selling them out of their family home in Sterling, Massachusetts. The Buttericks' monthly magazine, the *Delineator*, launched in 1873, would become the leading women's fashion magazine in the country by the turn of the century.

By 1876, E. Butterick & Co. had one hundred branch offices and one thousand agencies throughout the United States and Canada. Butterick's survives as a company to this day.

of her sister and husband, to enable women to adapt her patterns to any size or figure they pleased.

It was the right idea, at the right time. More women than ever were taking up dressmaking with the advent of the first sewing machines (see page 97). (Madame Demorest invented her own sewing machine, too: one that ran backward as well as forward.) The new railroads could rush paper patterns, selling for anywhere from twelve cents to a dollar apiece, to women all over the United States. "Elegantly trimmed" dresses, skirts, blouses; men's and children's and infants' clothing—there was a pattern for everything. Dress sections—bodices, sleeves, mantles, basques—were sold separately so they could be used on whatever outfits women liked.

Soon Nell was traveling frequently to Europe and mailing the latest trends and fashions from Paris and London back to her sister Kate, now working as her chief stylist. She continued to design and make dresses herself, provided custom-made patterns at her top clients' request, and added lines of lingerie, perfume, cosmetics, and sewing notions. She and her husband opened some three hundred satellite stores, "Madame Demorest's Magasins des Modes," and employed over 1,500 sales agents—almost all of them women—in cities around the United States, as well as in Canada, Europe, and Cuba.

For two extremely moral people, the Demorests possessed some *Mad Men* chops when it came to self-promotion. William Demorest started a series of immensely popular quarterly, then monthly, magazines that promoted women's suffrage, civil rights, and temperance, alongside articles on fashion, plates of dress patterns, and sample paper patterns stapled into the binding. Nell published the annual *Madame Demorest's What to Wear and How to Make It*, along with quarterly catalogues. She was always represented at the big European shows, installed a huge exhibit at the Philadelphia Centennial Exhibition, and created a wedding trousseau and wardrobe for Lavinia Warren, whose marriage to her fellow circus "midget," "General" Tom Thumb, was promoted by P. T. Barnum into a worldwide sensation. Her work drew praise everywhere.

"What Madame Demorest says is supreme law in the fashion realm of this country," one rival admitted.

By 1876, Madame Demorest was selling over three million paper patterns annually in the United States and Europe, but her reign was already giving way to that of another industrious couple. Ebenezer and Ellen Augusta Pollard Butterick copied many of the Demorests' promotion and distribution methods—even starting their own magazine, the *Delineator*—and patented their patterns, something the Demorests had neglected to do. More important, the Buttericks sold their patterns *already* fitted to many different sizes; no need to use Nell Demorest's drafting system, no matter how brilliant it was.

The Demorests sold off their business in 1887. They had made a great deal of money and seemed just as content to devote themselves to their many social causes. In these, they had always been consistent when it came to business. Almost alone among large American companies in the nineteenth century, Mme. Demorest's hired numerous blacks as well as whites for all business operations and treated them equally. If Nell's clients objected, she told them they could go elsewhere. If her white employees objected, she fired them. She was, as in fashion, always a fearless trendsetter.

COPPER-RIVETED JEANS

They are an American icon, invented and marketed on the Western frontier by a pair of immigrants. A work garment turned fashion statement, jeans—formerly "blue jeans"—would become the favorite clothing of people all over the world, in every walk of life, though they had the humblest of origins.

Jacob Youphes started life as a subject of the czar, a tailor of German-Jewish extraction born in Riga, the capital of Latvia, then a province of the Russian Empire. Emigrating to the Lower East Side of New York at the age of twenty-three, he changed his name to Jacob Davis and led a peripatetic existence as a journeyman tailor, living in Maine, California, British Columbia, then Nevada. Along the way he also tried his hand at panning for gold, selling tobacco and foodstuffs, and finally investing his life's savings in a Reno brewery. The business went belly up, leaving Jacob to fall back on his tailoring to support his wife and six children.

A lazy husband and his determined wife would give Davis a whole new opportunity. One day in 1870, a local woman asked him to make a pair of pants for her husband. It seemed that she wanted him to chop some wood, but he claimed to have no pants fit for the job. His wife gave Davis three dollars up front and told him to make the pants "strong" for her man, variously described as "very large" and "enormous."

She had come to the right tailor. Davis of late had been specializing in making wagon covers and tents out of ten-ounce, white cotton duck cloth for a prosperous merchant in San Francisco named Levi Strauss. Cotton duck cloth was an incredibly tough fabric, but Davis also happened to have on hand some copper rivets, which he used to make straps on horse blankets for teamsters to use. He added them to the pants pockets for his corpulent client, just to make sure they wouldn't tear.

The pants were an immediate success, and soon Davis was inundated with orders for his copper-riveted "waist overalls," making some of them out of nine-ounce, blue denim material that he also got from Strauss. Western work tended to be hard outdoor work—digging mine shafts, breaking horses, driving cattle, pounding fence posts, standing for hours in freezing streams to fish or to sift for gold—and pants needed to be tough. The copper rivets also proved ideal, holding together pockets that were often stuffed with work tools.

Among his many talents, Davis was an inveterate inventor, and he saw the potential of what he had at once. But his English was still limited, and his initial patent application for the jeans was denied. Afraid that he would lose his idea altogether, something that had happened with an earlier invention, and hard-pressed to raise the sixty-eight dollars necessary for a new application, he wrote to his San Francisco contractor, who quickly agreed to share the patent with him.

Levi Strauss, also a German-Jewish immigrant, had emigrated from Bavaria the year before Davis came over. He had also changed his name, from Loeb to Levi, and had also moved about the country, working for his family's dry goods business, first in New York City, then in Louisville, Kentucky, and California.

Strauss's command of English and the American legal system seems to have been considerably more advanced than Davis's at this time, and a less scrupulous man might well have seized the opportunity to file Davis's patent for himself, and made a killing. Instead, Strauss invited Davis to San Francisco, where their partnership progressed rapidly, and Strauss soon shifted the whole orientation of his business from tents to pants. These were sewn first by women in their homes but then in a factory where Davis supervised some 450 workers. Their sole material soon became denim, which was shipped in by Strauss from the high-quality Amoskeag textile mills in Manchester, New Hampshire. Unlike cotton duck, denim becomes softer and more flexible with repeated wearing and washing.

By the time Davis and Strauss died, decades later, both were wealthy men. Leadership of their company passed to some of Strauss's nephews and one of Davis's sons, and both families are still making jeans to this day. But jeans remained pants worn mostly by working men and little boys until the 1940s, when *Life* magazine photographs of women students at

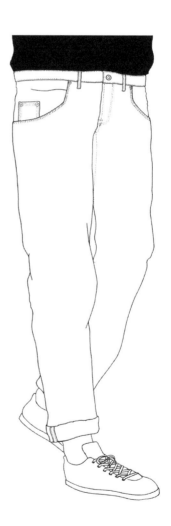

Pants for all seasons:
copper-riveted jeans today.

Denim is a sturdy, cotton, warp-faced, twill textile. Traditionally, its warp threads were dyed in indigo and its weft threads were not, which is why most jeans look different inside and out.

The word *denim* is derived from *serge de Nimes*, or "fabric of Nimes." The word *jeans* derives from the French word for Genoa, Italy—*Genes*—where the first denim pants were made.

The oldest extant pair of Levi's jeans is from the 1880s and was found in an abandoned mine in Colorado. The Levi Strauss Company paid $25,000 for them, and collectors now search old mines for more discarded jeans from the nineteenth century.

In the 1870s, Davis added a double arch of stitching to the back of Levi's jeans' pockets to distinguish them from imitators. It is the oldest extant apparel trademark in America today.

The official name of Levi's was changed from "waist overalls" to "jeans" in 1960.

Radcliffe wearing jeans shocked much of America. Soon a pair of cinematic rebels, Marlon Brando and James Dean, made jeans cool for the first time. It wasn't long before jeans in every possible shape, size, color, and style—ripped, faded, stonewashed, black, blue, white, hip hugging, relaxed, flattering, comfortable, skinny, fat, boot, and bell-bottom—were sold everywhere.

Today, over 1.2 billion pairs of denim jeans are sold around the world every year—a $56 billion business. The average American owns seven pairs of jeans.

LIBERATED BY THE BRA

Mary Phelps "Polly" Jacob received the first patent for the modern brassiere in 1914. It was a propitious time for an unconventional woman, and her invention.

Variations on bras had existed for centuries in cultures all around the world. Major movements for "rational clothing" among women had sprung up in England and the United States in the 1800s, and the *bandeau* had begun to make real inroads in French fashion, always a trendsetter. Nonetheless, the whalebone (really "baleen"; see page 159) and metal corset had reached new lows of constrictive cruelty by the turn of the century. Jacob, daughter of a blueblood American family that traced its roots back to the Mayflower and to steamboat inventor Robert Fulton, was sick and tired of this underwear armor that tortured her buxom figure. Preparing for a friend's debutante ball one night in 1910 when she was still just nineteen, she had her maid bring her two pocket handkerchiefs and some pink ribbon. Together they sewed up what Jacob would call the "Backless Brassiere."

The bra was an instant success, allowing Jacob to move around the dance floor with unprecedented grace and freedom. All her friends wanted one, and a family lawyer persuaded her to file for a patent, where she described her invention as "well adapted to women of different size" and "so efficient that it may be worn by persons engaged in violent exercises like tennis."

By 1920, Jacob had founded the Fashion Form Brassiere Company, after having to make the formal declaration that she was a married woman using funds separate from her husband's bank account. She opened a two-woman factory on Boston's Washington Street and managed to sell a few hundred bras to local department stores, but the business never really took off—perhaps in part because she also used her shop as a trysting place with an extramarital lover. Jacob happily sold her patent for $1,500 to the Warner Brothers Corset Company in Connecticut, which soon discontinued her model but used the patent to make an estimated $15 million.

Jacob never looked back. After losing her first husband to alcoholism, she married the wealthy, erratic, and mystical dilettante Harry Crosby, a man six years her younger, and moved with him to France, where she changed her name to "Caresse Crosby." The

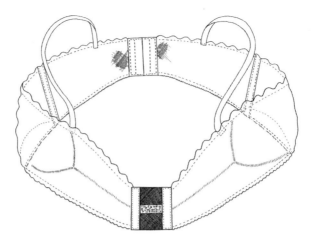

Bohemian Caresse Crosby's patented bra, first invented in 1910, freed women from the cruel constrictions of the corset and allowed them to move freely.

two of them put the "Lost" into the Lost Generation with a spree of extraordinary libertinage—in between supporting most of the leading artists of their time.

"I can't say the bra will ever take as great place in history as the steamboat, but I did invent it," she would write in her memoir, where it was, unsurprisingly, a bit of a footnote.

An altogether more serious couturier was Ida "Itel" Rosenthal, née Kaganovich, born near Minsk in 1886, who like so many young Jewish women of that time and place was taught to sew at a young age by her mother. Fleeing the Russian Empire with her fiancé, William Rosenthal, after their involvement in the failed revolution of 1905, she went to work for herself as a dressmaker in their small Hoboken apartment. There she raised and supported two children and her husband as he struggled through different tuberculosis sanatoriums.

Over the course of fourteen years, she built up a two-story sweatshop, with fifteen workers in their house, then moved the business across the Hudson to West Harlem. There, in 1921, she partnered with Enid Bissett, a former vaudevillian who was running an upscale dress shop on West Fifty-Seventh Street. Neither liked the way bras of the time fit or how they spoiled the way their dresses looked on customers.

Corsets were now fully on the way out—as chairman of the War Industries Board, Bernard Baruch had implored women to give up their corsets and had claimed their dedication to the war effort saved twenty-eight thousand tons of steel, or enough to make a battleship—but the "flapper" look that replaced them in the 1920s proved to be just one more example of men deciding what they thought women should look like at any given moment.

"It was a sad story. Women wore those flat things like bandages," Rosenthal later told a reporter. "A towel with hooks in the back. And the companies used to advertise, 'Look like your brother.' Well, that's not possible. Why fight nature?"

To drive home their point, the two women offered a brand of bra called "Maidenform," at Bissett's suggestion, in direct contrast to a popular flapper bra known, believe it or not, as "Boyishform." Bissett and Rosenthal began to make modern bras out of soft-knit mesh, with pockets and uplift. They were so good, in contrast to the rather crude, bosom-crushing garments that went before them, that soon the women's dress customers were asking for extra bras to go with their outfits. During the endless hours of enforced rest in his sanatoriums, William Rosenthal had spent hours sculpting women's bodies with clay—something he also did, almost compulsively, on beaches with sand—and now he was able to make bras for all shapes and sizes.

Itel Rosenthal handled all the rest of the business, eventually buying out Bissett when she wanted to retire and bringing her daughter, Beatrice, and son-in-law, Joseph Coleman, into the company. She made them both learn the business from the ground up. Beatrice proved as capable an administrator as her mom, while Joseph was smart enough to snap up an ad campaign that another woman pioneer, Mary Filius, conceived, the famous "I dreamed I . . . last night, in my Maidenform bra," a racy (for the time) series of magazine ads that stressed the product's comfort and ease in wearing.

If it wasn't the steamboat, a Maidenform bra was bought by one in every five American women, and the company reported $100 million in sales by 1980. It employed five thousand people at twenty-eight American locations and sold product in 115 countries. The company stayed in the extended Rosenthal family until 1998.

"Caresse" and Harry Crosby bought Henri Cartier-Bresson his first camera. They also founded Black Sun Press, which published seminal works by William Faulkner, James Joyce, Ernest Hemingway, Archibald MacLeish, Ezra Pound, D. H. Lawrence, Hart Crane, T. S. Eliot, Henry Miller, and others in affordable paperback editions.

Following Harry Crosby's death in a suicide pact with a lover, Caresse kept Black Sun going for a few more years; wrote pornography with Anaïs Nin; took on another younger, drunken, troubled husband; ran artistic and literary salons; tried to start a world peace center; and published *Portfolio: An International Quarterly*, which published Man Ray, Gwendolyn Brooks, and Max Ernst and artwork by Picasso and Dalí.

Itel Rosenthal started as a seamstress in America with a Singer Sewing Machine (see page 97) that she bought on an installment plan.

It was William Rosenthal who, for his wife's company, came up with the modern lettered sizes for bra cups.

During World War II, Maidenform made bras for the Women's Army Corps (WAC), parachutes for the US Army Air Force, and vests for messenger pigeons.

IT'S NOT DRY
DRY CLEANING

How to keep one's clothing clean without shrinking or ruining it is a concern that goes back at least to the Romans and probably the ancient Greeks. The Romans hit upon the idea of weaving what we call fuller's earth—claylike earth, with traces of alkaline and other chemicals—right into their woolen togas. Togas were also cleaned in part with combinations of lye and ammonia. The ammonia was collected from the urine of farm animals—or friends, Romans, and countrymen. Outside public latrines were special pots, thoughtfully left there by the local *fullonicae*, the laundries that were often the biggest employer in any district and were owned by wealthy and prominent men. Fullers' guilds soon abounded, and the government taxed the collection of urine, which provided a steady stream of income.

Over the next two thousand years or so, people evolved to the point where they didn't necessarily have to dunk their finest garments in their own pee. By the late seventeenth century, launderers were using kerosene—also known as "white naphtha" or "coal oil"—spirits of turpentine, and related substances, all of which were very good at getting out oil-based stains, and exploding into flames. Clothes were simply washed in vats of this stuff, then hung out to dry, leaving your clothes smelling like a barbeque, among other things. Dry-cleaning plants were considered so dangerous that almost no one would insure them, and dry cleaners began the custom of accepting your dirty clothes in little shops, then sending them off to plants far from where anyone lived.

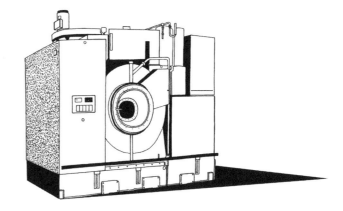

The modern dry-cleaning machine, a miracle of recycling, with its closed system capturing most of the solvent it uses.

In 1821, a thirty-year-old New Yorker named Thomas L. Jennings became the first African American ever to be granted a US patent, for a new dry-cleaning process he called "dry scouring." It was no small feat for a black man to make his way in the New York City of the time, where slavery was still legal until 1827 and people of color were frequently subjected to abuse and violence. The very fact of his winning a patent raised a hue and cry, for it was against the law for a "slave" to own any such thing. What the white supremacists didn't know was that Jennings had been born to free parents. He became a tailor and was apparently so talented that he had his own shop on Church Street, in Lower Manhattan, catering to an upper-class clientele.

Noticing that many of his customers were irate over stains that ruined their clothes, Jennings came up with dry scouring. Little is known about this process—its secret was lost in a fire in 1836—but it must have worked. Jennings reportedly became rich. He bought his wife and her family out of bondage and devoted most of the money he made to the fight for freedom and civil rights.

Thomas and Elizabeth Jennings's daughter, also Elizabeth, would anticipate Rosa Parks by over a hundred years, when she was forcibly ejected from a horse-drawn Manhattan streetcar—after a prolonged physical struggle—by a conductor and a policeman in 1854. Elizabeth, a schoolteacher on her way to play the organ at her church when she was ejected, sued the Third Avenue Railroad Company, and won $272.50 in costs and damages—and more important, the desegregation of all the company's streetcars. Her case, funded by her father, was handled by future US president Chester Allan Arthur. Elizabeth Jennings would go on to a long career in education, starting New York's first kindergarten for black children in her home.

Gasoline and other inflammables would join kerosene as "cleansers" by the turn of the century. But by the 1920s, William Joseph Stoddard, a dry cleaner in Atlanta, had combined with Lloyd E. Jackson, a scientist at the Mellon Institute, to come up with "Stoddard Solvent," a somewhat less volatile, petroleum-based "white spirit" that would soon become the leading dry-cleaning solvent and would remain so for over thirty years.

The dry-cleaning industry, though, had also begun to diversify into chlorinated solvents—thanks in part to government research efforts aimed at producing poison gas during World War I. By the 1940s, one of the new solvents, tetrachloroethylene, better known as perchloroethylene, or "perc," was being used more and more. Perc proved to be stable, much less prone to explode in flames, and easy on your silk underwear.

It is also, alas, a carcinogen. This does not mean you are in any danger simply by wearing dry-cleaned clothes, mostly because all but infinitesimal traces of perc are removed from them before they come back to you.

But the process, like all dry-cleaning processes, is not dry. Your clothes were and are cleaned in what is usually a sort of combination industrial washing-drying machine. First

THE GENIUS DETAILS

Surviving photographs of Thomas Jennings's dry-scouring machine show it resembling an iron presser.

Clothes are usually dry-cleaned at 86 degrees Fahrenheit and air-dried at 140 to 145 degrees Fahrenheit.

Modern machines recover 99.99 percent of all solvent used.

Between 2006 and 2011, the dry-cleaning industry averaged annual revenues of $7.5 billion and included twenty-two thousand businesses, employing some 150,000 workers.

Extensive research involving perc was done during World War I. At temperatures over 600 degrees Fahrenheit, it oxidizes into the extremely poisonous gas phosgene, the most effective of the chemical weapons utilized in the war, which killed an estimated eighty-five thousand troops.

they're washed in one chamber full of liquid perc, along with water, a little detergent, and other specialized cleaning solvents. They are then air-dried in a second chamber, with the perc evaporating into vapors that are condensed and collected, while—nowadays—computer sensors confirm that it has been removed.

For decades, this was all done in a "vented" system, in which fumes were released freely into the air—or into the lungs of laundry workers. All dry-cleaning plants now use "closed" systems, in which nearly all the perc solvent is recaptured and reused over and over again as it changes from a liquid, to a gas, to a solid, and back.

Even so, concerns over perc causing cancers in exposed workers, and of contaminated wastewater slipping into drinking supplies, led California to ban perc by 2023, setting off a scramble by dry cleaners to find a replacement. A wide variety are already being tried, ranging from glycol ethers to hydrocarbon, liquid silicone to liquid carbon dioxide, and modified hydrocarbon blends to brominated solvents. All have their advantages and disadvantages when it comes to both the environment and getting out those pesky, congealed blood splatter stains. Maybe if someone could turn up Thomas Jennings's old patent. . . .

SNEAKING AROUND
THE ATHLETIC SHOE

Believe it or not, there was a time, not very long ago, when most American adults did *not* wear rubber-soled athletic shoes whenever they could. Yet the origins of the running shoe can be traced back to well before the Civil War.

It was a craze for betting on foot races, in late eighteenth-century England, that created the desire for an athletic shoe in the first place. Everybody, it seemed, was running: men, women, young people, old people, fit people, fat people, rich people, poor people. The bets could be quite elaborate, such as whether one could eat a chicken while running a certain course. Early athletic shoes were developed for these runners, but they tended to be made of leather, and they stretched out of shape when wet. (There was also a limited need for them, as some runners in these competitions ran completely naked.)

By the 1830s, the Liverpool Rubber Company was selling "Plimsoll shoes"—light, rubber-soled shoes that were generally used for the beach, and later for children's physical education classes. This would set up the enduring conflict in athletic shoes:

Should they be as light and comfortable as walking on rubber, or as tough and durable as leather?

One Walt Webster of New York tried to split the difference by patenting a process in 1832, whereby a rubber sole was attached to a leather boot or shoe. But Webster was ahead of his time. Natural rubber tends to turn stiff and brittle in extreme cold, melt into a shapeless mass in the heat, and exude an awful stink when it absorbs grease, oil, or acid. It would not be until well into the next decade that Charles

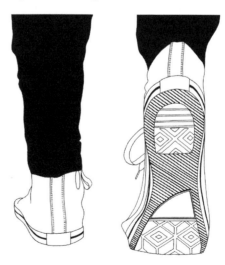

Chuck Taylors: the basketball shoe that became a fashion statement.

Goodyear perfected a process of vulcanizing rubber that made it commercially viable, and it was not until the 1890s that bicycle companies began making rubber athletic shoes.

During World War I, the giant conglomerate U.S. Rubber consolidated some thirty existing brands into the single rubber-soled, canvas-topped sneaker known as "Keds"—soon to be an American classic. But Keds remained a flat, casual shoe, offering little support. Real athletes were still wearing leather or canvas shoes, with metal cleats or actual spikes for traction (or spiking opponents).

This had started to change in 1907, when the sporting goods giant A. G. Spalding began to design shoes specifically made for playing tennis, and then basketball. A major breakthrough came when a small Boston firm called Converse hired a twenty-year-old former high school basketball player from Indiana named Chuck Taylor to sell its shoes. Taylor had walked into the company's Chicago office in 1921, complaining that its basketball shoes made his feet hurt. He stayed on to help redesign Converse shoes, with their distinctive "high-top" look to increase support and flexibility, and with a "protective" ankle patch that would soon bear his name. Irresistibly friendly and charismatic, Taylor traveled around the country for decades in his white Cadillac with a trunk full of "Chucks," holding basketball clinics and tirelessly promoting both the brand and the game. He lobbied constantly for basketball to become an Olympic sport, and when it did, in 1936, the US basketball team was entirely decked out in "Chuck Taylor All-Stars."

Yet it was the transition of running to a mass activity that really allowed the athletic shoe to come into its own. Bill Bowerman, the legendary men's track coach at the University of Oregon, stumbled onto the idea in 1962 when he visited a colleague, New Zealand Olympic coach Arthur Lydiard, who had developed a regimen of "jogging" as a fitness program for people of all ages and physical conditions. Fascinated by this, Bowerman returned to the United States, developed a model running program, and, with the help of a cardiologist, published a book on running as exercise.

He also began to experiment with making shoes on his own, something that proved over time to be a lucrative, if dangerous, obsession. He created a cushion heel wedge and did everything he could to strip weight from his shoes. Famously, he even used his wife's waffle iron to create a new "waffle tread" that proved wildly popular, its prints resembling the footsteps the astronauts had left on the moon. Bowerman's smartest business decision, though, was no doubt a handshake agreement to start a shoe business with a former miler of his, turned accountant and assistant business administration professor, by the name of Phil Knight. The company they founded, first called Blue Ribbon Sports and then Nike, would become one of the most successful—and controversial—sporting goods companies in the world.

Other shoe companies in America and abroad would continue to push the technology of the humble sneaker forward. New Balance, once a small Boston arch support company, introduced the first ripple-soled athletic shoe and the first scientifically tested running shoes early in the 1960s. Spikes became shorter, made out of ceramics and light alloys. NASA's development of "blow rubber molding" to air-cushion helmets and shoes would be adapted by Nike into its "air" basketball shoes, mitigating the pressure felt on heel strikes. Heel wedges, midsole wedges, orthotics, and shoes of different widths would be customized to fit all kinds of feet in all kinds of conditions.

Today, nearly one in every six Americans runs for exercise—nearly one-tenth of them at least once a week. As debates continue over *how* exactly we should run—what parts of our feet should hit the ground first, and how hard—the running shoe will continue to evolve.

THE GENIUS DETAILS

"Plimsolls" were named after the "Plimsoll Line" on boats, showing where the hull would meet the water. Anything above it and the foot would get wet, as well as the sailors.

New Balance was founded in Boston as an arch support company by British immigrant William J. Riley. Riley always kept a chicken foot on his desk to illustrate what he felt was the "perfect balance" of the three-toed claw.

Promoting his product, Nike's Phil Knight announced that four of the first seven finishers in one of the 1972 US Olympic trials were wearing Nikes. He failed to mention that the first three finishers wore Adidas shoes.

Nike's worldwide revenues were reported in excess of $24.1 billion by 2012, and it directly employs forty-four thousand individuals in the United States. Millions more work for pennies an hour in the seven hundred factories it owns around the world.

Nearly all running shoes are made outside the United States today. New Balance is the leading domestic manufacturer, making one-quarter of its shoes in America.

THE DISHWASHER AND THE DIAPER COVER

Catharine Esther Beecher, a groundbreaking educator, writer, and lecturer who traveled all over the United States, did not believe that women should have the right to vote or take direct part in "civil and political concerns." Beecher, who took over running the household of her family at sixteen, after her mother's death, and saw to the welfare and education of her remarkable younger siblings, believed that women had a much more important role already. They were moral guardians of society and administrators of the home, which was "the aptest earthly illustration of the heavenly kingdom."

Women might work as teachers, or seamstresses, or midwives (or writers, such as herself and her famous younger sister and sometime collaborator, Harriet Beecher Stowe, whose *Uncle Tom's Cabin* became a worldwide bestseller). But they should not expect to enter business or the professions. Regressive as such opinions may seem to us today, Beecher did hold the radical view for her time that women should marry whomever they liked, or not marry at all—as Beecher never did, following the tragic early death of her fiancé.

Beecher also found the education of women in the "domestic arts" to be woefully informal and arbitrary, something she set out to correct, starting with her 1841 manual *A Treatise on Domestic Economy*. Reprinted every year through 1856, it would become almost as well known in America as her sister's work. Containing instructions for "everything from the building of a house to the setting of a table," it marked the start of "home economics" as a real discipline. As Beecher declared, "The care of a house, the conduct of a home, the management of children, the instruction and government of servants, are as deserving of scientific treatment and scientific professors and lectureships as are the care of farms, the management of manure and crops, and the raising and care of stock." In a day when running a household with or without servants often entailed ceaseless toil and backbreaking drudgery, what Beecher was attempting was a sort of feminism of the deed, in the sphere where women of her time—like it or not—would spend most of their lives.

A phalanx of Beecher's sisters were already hard at work, devising ways to make that labor a little easier. If there are relatively few women in the ranks of the great commercial inventors, it is largely because they were denied, by law or custom, the opportunity to do such work. But little by little, they were busily changing their world—and, in so doing, building their own opportunities for the future.

Sometimes, meanwhile, what they invented in the home became business—even big business.

The earliest such woman we know of was Sybilla Masters, who in 1715 became the first recorded woman inventor in American history. Dame Masters, a native of Bermuda then living in Philadelphia, had to have her patent for "Cleansing Curing and Refining of Indian Corn Growing in the Plantations" issued in the name of her husband, Thomas, by the British courts, as they would not recognize a woman inventor. Masters's mill consisted of a wooden gear and shaft, motored by a donkey, and resembled "something after the manner of a musical box." She would later issue a second patent through her husband, "Working and Weaving in a New Method, Palmetta Chip and Straw for Hats and Bonnets and other Improvements of that Ware."

Josephine Garis Cochran was a forty-four-year-old, heavily indebted widow and mother of two living in Shelby County, Illinois, in 1883, when she decided she was tired of the servants chipping her heirloom china. She set to work in a shed behind her house to invent a dishwashing machine, assisted by a mechanic named George Butters.

Dishwashers had been attempted before, but none had proved practical. Cochran, the daughter of a steamboat inventor, designed a remarkably modern device, with measured wire compartments customized for plates, cups, and saucers. All were then placed in a wheel that was laid flat inside a copper boiler. While the wheel was cranked by hand, hot, soapy water squirted up from the bottom of the boiler and over the dishes. It was the first

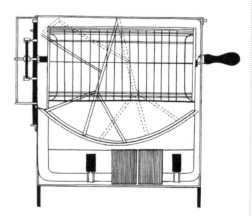

Josephine Garis Cochran's revolutionary 1886 dishwasher, the first to use water pressure (from the pumps along the bottom) instead of scrubbers to clean the dishes, which were held on the rack near the top and turned by the lever on the right.

THE GENIUS DETAILS

Sara Elizabeth Goode, née Jacobs, who may have been born as a slave in 1850, became on July 14, 1885, the first African American woman ever to earn a patent, for her "cabinet bed," a complete desk that folded up into a bed.

Martha Helen Kostyra reinvented herself as "Martha Stewart" and described her career as "just a different way of looking at things that women had been doing for centuries . . . that was more considered drudgery than joy."

By the turn into the twentieth century, "home economics" was being taught widely in American high schools and in some colleges and universities. Eleanor Roosevelt would call home ec "the most important part of the university, for it concerns the homes of the people of this country."

Today 75 percent of all US households have an automatic dishwasher. Due to increased efficiency, Americans use just one gallon of water a day, per person, on dishwashers, and dishwasher water is just 1.4 percent of all home water use.

such machine to use water pressure to clean instead of scrubbers.

Cochran made dishwashers for her friends, earned a patent for her design in 1886, opened a factory, and went into business. Her design won a top prize at the 1893 World's Columbian Exposition in Chicago for "best mechanical construction, durability and adaptation to its line of work." She sold nine models to eateries at the fair.

Few American homes would have the capacity for hot water to handle a dishwasher until the 1950s, and the price of $150 a machine was high for the nineteenth century, but Cochran was able to sell her machine to restaurants, hotels, and hospitals, which liked its sanitizing hot rinse. She kept improving her invention, substituting an electric motor for the hand crank, adding a drainage hose, and getting the racks to revolve. It was no easy task.

"I couldn't get men to do the things I wanted in my way, until they had tried and failed in their own," Cochran told a reporter late in life, while still on the job. "And that was costly for me. They knew I knew nothing, academically, about mechanics, and they insisted on having their own way with my invention until they convinced themselves my way was the better, no matter how I had arrived at it."

Cochran managed the company herself for the rest of her life, made her own visits to pitch her machines to hotels, and got her Shelbyville neighbors to invest in it. Three years after her death in 1913, "Cochran's Crescent Washing Machine" was acquired by the company that became KitchenAid, which would in turn be absorbed by Whirlpool.

Marion O'Brien Donovan graduated from college with a BA in English shortly before World War II and went on to stints as a beauty editor at *Harper's Bazaar* and *Vogue*, a résumé that qualified her to . . . get married, have a daughter, and start changing diapers.

Like countless mothers of the time, she was frustrated by how much laundry cotton diapers required—and how they still didn't keep baby from soiling bedsheets and other materials. There were already rubber pants on the market, but they tended to cause diaper rash and pinch the infant's skin.

Marion Donovan set to work with her sewing machine (see page 97) and a piece of shower curtain. Three years, countless shower curtains, and one more baby later, she had a comfortable, waterproof diaper cover, complete with snap fasteners, to avoid sticking baby with a safety pin. The final product was made of nylon parachute cloth, which reduced diaper rash. She called it the "Boater" because it reminded her of a boat. Manufacturers thought it reminded them of nothing they wanted to produce. Diapers? Why should men worry about that?

Setting out on her own, Donovan managed to sell her Boaters to Saks Fifth Avenue

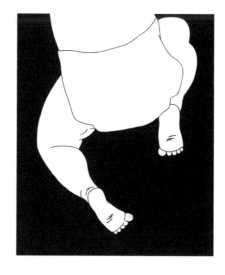

Marion O'Brien Donovan not only invented the 1940s plastic diaper cover depicted here, but also a protoype of the disposable diaper.

in 1949 and had won four patents for their design by 1951. It was an instant smash, and Donovan quickly sold the patents to the Keko Corporation for a reported $1 million.

She could easily have retired on her laurels, but instead, according to writer Kate Kelly, "She used that money to fund other inventions; her goal was always to create products that made life convenient and more organized." Catharine Beecher would've been pleased.

Donovan went on to earn sixteen more patents for such items as two-ply dental floss that eliminated the need to wrap floss around one's fingers, a towel dispenser, a hosiery clamp, a facial tissue box, and a closet organizer she called "The Big Hangup." She went back to school, graduating with a master's degree in architecture from Yale in 1958, at age 41, one of only three women in her class. She worked in a Connecticut firm, using what she had learned to design her own home ("Everything from the building of a house . . ."), and later became a corporate consultant on home product development.

Still, Donovan didn't turn her back on diapers, coming up with an idea for a disposable diaper, one that would absorb water and pull it away from the skin. Once again, no one was interested—until ten years later, in 1961, when Victor Mills tried his hand at inventing just such a disposable diaper for Proctor & Gamble. It was called "Pampers."

FROM LIQUID PAPER TO KEVLAR

B y 1910, women were awarded less than 1 percent of all patents issued in the United States (as opposed to 7.5 percent of them awarded in 2012). But despite all the barriers they faced, women inventors did emerge into the workplace, where they were, predictably, often dubbed "Lady Edisons" with the usual condescension of male reporters.

The first Lady Edison was Margaret E. "Mattie" Knight. Working with her brothers in the massive Amoskeag Mill, in Manchester, New Hampshire, in 1850, she saw a boy stabbed by a spindle that broke loose from its cotton loom. Knight, just twelve years old, quickly conceived of a device that would automatically shut off the loom if something went wrong. Her invention was developed and put into use—supposedly by mills all over the country—though she never saw a dime from it.

By the time she was thirty, she was working in a paper bag company in Springfield, Massachusetts, pursuing the rough, seminomadic existence of a poor, unmarried woman in industrial New England. But Mattie was always fascinated by machines and ways to improve things. She understood that the bags—which were more like big paper envelopes at the time—would be infinitely more useful if they had flat bottoms that would allow them to be stood up and more easily packed.

Knight designed a machine that would fold the bags and glue their bottoms into place. She built her own wooden prototype of the machine, then asked a mechanic at her plant, Charles Annan, to help her build an iron prototype for her patent application. Annan agreed—then claimed the idea was his. This time, Knight was an adult woman, and she hauled Annan into court.

Knight presented witnesses, years of drawings, and that wooden prototype to prove that the design was rightfully hers. Annan simply claimed that a woman could not possibly have invented such a device. It was an argument that had often prevailed in the past—but this time Knight won, getting her patent in 1871. She was one of the very first American women to win a patent in her own name (the very first was Hannah Wilkinson Slater, who invented a type of two-ply sewing thread back in 1793).

Paper bags, now looking much as they do today, took off. Knight founded the Eastern Paper Bag Company to produce them, though she was confounded when her male workers refused to take orders from her, unwilling to believe a woman knew how machinery worked, even if she had worked machines for most of her life. But she raked in royalties and went on to win twenty-seven patents in her lifetime for inventing another machine that cut out shoe soles, a "dress shield" to protect clothes from sweat stains, a rotary engine, an internal combustion engine, an automatic boring drill, and a window frame and sash. A few months before her death, the *New York Times* reported that Knight, "at the age of seventy, is working 20 hours a day on her 89th invention."

This was not quite accurate; Knight was already seventy-five. Her one regret? "I'm only sorry I couldn't have had as good a chance as a boy, and have been to my trade regularly."

One of the most incredible stories involved Bette Clair McMurray Nesmith (later Bette Graham), a twenty-seven-year-old executive secretary at the Texas Bank and Trust in Dallas in 1951. Nesmith, a divorced mother who had started secretarial school when she was seventeen, was exasperated to find that she could no longer make a simple erasure correction of typing mistakes on the new IBM electric typewriters, which used a carbon-based film.

Inspired by watching painters work on the bank windows for the holidays, Nesmith—a professional artist herself—came up with her own mixture of white, water-based tempera paint to cover over her typing mistakes. By 1956, "Mistake Out" was in heavy demand, and Nesmith was selling it out of her North Dallas home with the help of her son, Michael—later a member of the TV pop sensation the Monkees and a Hollywood producer. By 1958 she had perfected her product, renamed it "Liquid Paper," and patented it.

Company sales were over one million units a year by 1967, and the former secretary now had her own automated production plant and corporate headquarters, despite heavy competition from the likes of "Wite Out" and "Papermate." The little black-and-white bottles were ubiquitous—icons of corporate ennui, with their labels so often altered to "Liquid ape" or "id Pap" by bored office workers.

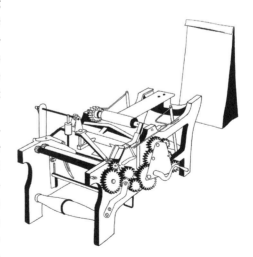

Mattie Knight's machine for making the modern paper bag, with its groundbreaking flat bottom. Today Americans use an estimated 10 billion paper grocery bags every year.

In 1975, Nesmith moved her company, which now employed two hundred people and sold twenty-five million bottles a year, into its new thirty-five-thousand-square-foot International Liquid Paper headquarters in Dallas. Stressing product quality and a decentralized decision-making process, she provided her employees with a fishpond, a library, and onsite child care.

Nesmith had hit the sweet spot of her profession's progression, the era between the manual typewriter and the advent of the personal computer, when male executives everywhere would miraculously gain the ability to type and corrections could be made with the push of a key. Perhaps understanding this, she sold her enterprise to Gillette for $47.5 million in 1979,

Bette Nesmith's original Mistake Out, renamed Liquid Paper: a revolution in covering up errors.

six months before her death. She left behind a major philanthropic foundation and the example of a woman whose inventiveness and fortitude took her to the top.

Stephanie Louise Kwolek, whose parents were both Polish immigrants, learned a love of sewing and fabrics from her mother. Kwolek briefly considered fashion as a career but instead obtained a degree in chemistry in 1946 from what was then the women's college of Carnegie-Mellon University. She took a job as a chemist with the DuPont Corporation, while still contemplating a career in medicine. There she found herself in fashion after all, researching the new polymers that were transforming American fabrics. Her specialty was finding new fibers that might be created at extremely low temperatures—zero to forty degrees Celsius—and would not melt or decompose at very high ones (above four hundred degrees Celsius).

In 1965, she discovered to her surprise that under certain circumstances, rodlike, aromatic polyamides of a high molecular weight would line up parallel to each other, forming liquid crystal solutions that could be spun into fibers of very high strength and stiffness. Usually, the sort of thin, cloudy liquid her first experiments produced would be tossed away, but this was unlike any polymer solution ever quite prepared before in a laboratory, and Kwolek asked technician Charles Smullen to try spinning it. It was so turbid that Smullen refused at first, convinced the turbidity must be caused by small particles that would plug up the tiny holes in the spinneret.

It spun all right, though—and produced a material that was stronger than nylon, which Kwolek had worked on before. Nylon, nothing: it was five times stronger than *steel*, pound for pound, *and* resistant to fire. Kwolek found the fibers could be made even stronger by heating them.

By 1971, she and her team at DuPont's Pioneering Research Laboratory in Wilmington, Delaware, had produced the sorts of strong, stiff fibers that could be made into Kevlar, the material of "bulletproof" vests—not to mention approximately two hundred other items, including firefighters' boots, cell phones, gloves, airplanes, boats, canoes, armored cars, tires, ropes, fiber-optic cables, tennis rackets, hockey sticks, and skis. It would also be used to build hurricane-resistant rooms and reinforce steel bridges.

Kwolek had produced fashion that, if it did not stop a runway, would stop a bullet. She received many honors and at least seventeen patents but no extra money from DuPont, which had her sign away her rights to Kevlar before making billions off the material. The lost windfall never seemed to bother her.

"I don't think there's anything like saving someone's life to bring you satisfaction and happiness," she said.

After retiring, Stephanie Kwolek spent her time tutoring young students in chemistry, especially girls. The week of her death in 2014, aged ninety, the one millionth Kevlar vest was produced. They have saved the lives of an estimated 3,000 police officers.

Dr. Patricia Era Bath grew up in an almost unimaginably different world from ours today, a living embodiment of how far African Americans have managed to push past the barriers imposed by the old Jim Crow system. She would overcome them all and invent a revolutionary new procedure and device for removing cataracts with lasers.

Born in Harlem in 1942, Dr. Bath heard stories from her father, Rupert—a Trinidadian immigrant who had worked as a merchant seaman and a newspaper columnist before becoming the first black motorman for the New York City subway—of his travels around the world. This whetted her appetite to see other lands, other places. From her mother, Ruth, a housewife and domestic who was descended from African slaves and Cherokee Indians, Dr. Bath got her first chemistry set and a love of reading. From both her parents, she got the idea that she could do anything she put her mind to.

Taking part in a summer cancer research project run by Yeshiva University and the Harlem Hospital Center when she was still a sixteen-year-old high school student, young Patricia developed a mathematical equation to predict the rate of growth of a cancer—work that so impressed one of the doctors running the program that he included it in a paper he presented at a conference in Washington, D.C.

The Kevlar vest, made of fibers created by Stephanie Kwolek, savior of thousands of police officers.

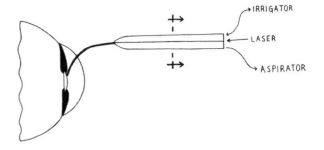

The patent for Dr. Bath's Laserphaco, the all-in-one laser device she invented to vaporize cataracts, and then wash and remove the cataract lens from the eye.

THE GENIUS DETAILS

Beulah Louise Henry, another "Lady Edison," was a direct descendant of Patrick Henry. She would come up with 110 inventions and win 48 patents. Henry's most successful patent was for an "Umbrella Runner Shield Attachment"— a snap-on cloth cover that allowed the owner to change her umbrella color to match her outfit.

Adm. Grace Murray Hopper served in the US Naval Reserve, reaching the rank of rear admiral, and her work in the Navy and at Remington Rand led to the development of some of the earliest computer languages, including COBOL, in 1959, which would become one of the most ubiquitous business languages in the world.

Cytogeneticist Barbara McClintock discovered chromosomal crossover and genetic recombination (cell division) in maize, a vital step forward in developing more productive generations of corn.

The publicity this garnered won her *Mademoiselle* magazine's 1960 Merit Award. Going on to med school at Howard University—her mother literally scrubbed floors to help her afford it—Dr. Bath became an intern at Harlem Hospital, then accepted a fellowship in ophthalmology at Columbia University. Her research at the Harlem Center showed that black people were twice as likely as whites in the general population to suffer blindness and eight times more likely to suffer from glaucoma. Concluding that too many black Americans were unable to afford proper medical care, Bath invented the discipline of "community ophthalmology," recruiting her Columbia professors to perform free eye surgery on needy Harlem residents, offer glaucoma screening and other preventive eye care, and provide glasses for children and the elderly—while she herself worked as an assistant surgeon. It was a volunteer effort that would be emulated on a global basis.

A long list of prestigious firsts, appointments, and trips to provide eye care around the world would follow—at one point she got married and gave birth to a daughter, all while completing a fellowship in corneal transplantation and keraprosthesis—but her most resounding contribution would be her invention of the Laserphaco Probe for cataract removal.

At the start of the 1980s, cataracts could still be extracted only through a difficult and extended surgical procedure that basically involved grinding them down—if they could be removed at all. Dr. Bath conceived of doing the job faster, more easily, and more safely with lasers.

Doing even basic research on this proved difficult. Most laser technology in the United States was reserved for military research. Dr. Bath had to travel to Berlin in 1981 to find an available laser, but after five years of work she had come up with the Laserphaco, a sort of "three-in-one" instrument consisting "of an optical laser fiber, surrounded by irrigation and aspiration [suction] tubes." The laser probe is inserted into a one-millimeter incision into the eye, where it vaporizes—"phacoblates"—the cataract and the lens matter, almost painlessly, and in just a few minutes. What remains of the cataract lens is then washed and sucked out of the eye by the irrigation and aspiration tubes, leaving it clean to insert a new lens.

By 1988, Dr, Bath had earned three patents on the device, making her the first African American doctor to earn a medical patent. She would use the royalties from them to fund the American Institute for the Prevention of Blindness and would go on improving the Laserphaco over the decades, as well as earning another patent for a method of dissolving cataracts with ultrasound technology. Today her invention is used around the world.

"The ability to restore sight is the ultimate reward," she would say.

THE GENIUS DETAILS

It also enabled her to become, in 1983 at the age of eighty-one, the first woman ever to win the Nobel Prize for Physiology or Medicine, unshared.

Dr. Ellen Ochoa became the world's first female Hispanic astronaut after earning her doctorate in electrical engineering from Stanford University. She used her education and experience on four space shuttle flights to become the coinventor of space-related optical recognition systems, optical inspection systems, and systems to remove "noise" from images. A classical flutist who also married and had two children, Ochoa was the second woman, and the first Hispanic American, to be named director of NASA's Johnson Space Center in Houston.

Bette Nesmith tried to sell her idea to IBM in 1956, but the company turned her down. She was fired from the bank she worked at in 1958 for accidentally signing the name of her Liquid Paper company to a bank business letter. Nesmith decided to go out on her own as an entrepreneur, even though she was selling only 2,000 bottles of her invention a year at the time. Liquid Paper remains a $120-million business today, even in the age of the computer.

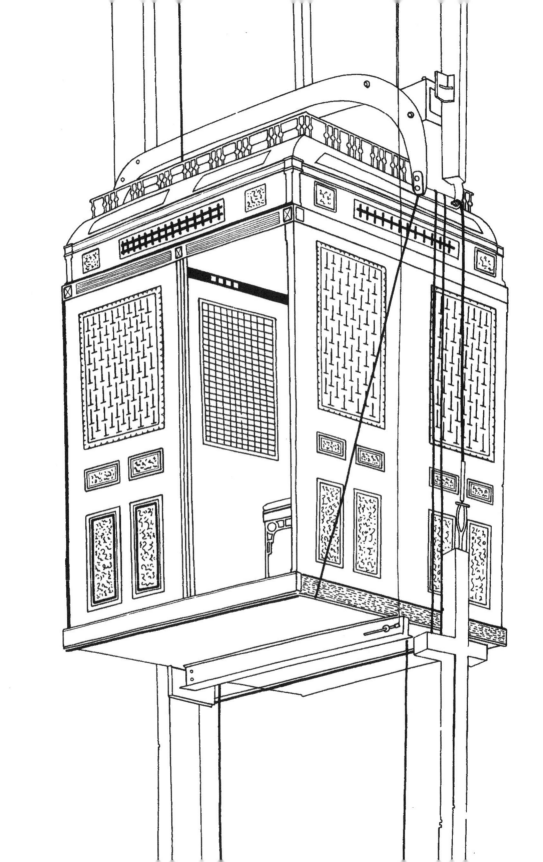

MR. OTIS'S SAFETY ELEVATOR

S ometimes it's the little things, the things we barely notice, that make much greater accomplishments possible. Combined, as it would be in the next generation, with those two other miraculous American inventions—the electric light and steel-frame construction—the safety elevator would make the skyscraper, and thus the modern city, possible. All thanks to one man's enduring faith.

He was, in many ways, the American Job, though many a frustrated inventor might have claimed the title. Like Job, Elisha Graves Otis was a pious, industrious man, respected by his friends and neighbors in rural Vermont; they would make him justice of the peace and elect him to the state legislature four times. He had left high school before graduating, as most Americans did in 1830, but, as his son would write, he "had no taste for a farmer's life." He drove a wagon instead, then started a gristmill, married a local young woman and had two sons with her, and built them all a house.

Then God started to test him, or so it must have seemed. The gristmill failed. Elisha turned it into a sawmill, but that failed, too. His wife died, leaving him with boys ages seven and two. Working in the bitter cold to make ends meet, Otis caught pneumonia and nearly perished as well. He married again and moved to Albany, New York, where he went to work in a bedstead company and invented a machine for turning out bed rails (sides) that enabled the company to increase its production from twelve beds a day to fifty. Rewarded with a $500 bonus, Otis started his own bedstead company, using a water-powered turbine he invented himself. Then the city of Albany diverted his water source, and his factory died. He tried to manufacture wooden carts, but that enterprise failed, too.

He invented a rotating, automatic bread oven; a steam plow; a new brake to let engineers stop locomotives faster. If one thing didn't work out, Elisha Otis remained mystifyingly confident something else would.

The Otis safety elevator would prove to be the vital part that made feasible the skyscrapers to come.

By 1873, there were over two thousand Otis elevators installed in America. Today there are some 1.7 million Otis elevators and 110,000 Otis escalators in operation in two hundred countries and territories around the world.

The year 1878 saw the first Otis hydraulic elevators.

Otis Brothers introduced directly connected, gear-driven, electric elevators in 1889, and in 1903, Otis set the industry standard with the gearless traction electric elevator, designed to outlive the building itself.

"He could invent, design, and construct a perfect working machine or improve anything to which he gave his mind, without recourse to any of the modern drafting methods," marveled his more practical-minded older son, Charles, whom he often drove to distraction. "He needed no assistance, asked no advice, consulted with no one, and never made much use of pen or pencil."

Moving on to Yonkers, Elisha made the great discovery of his life when he was hired to turn one more sawmill into one more bedstead factory. Looking to move some heavy debris to an upper story of the building, he constructed a hoist, or "elevator." This was nothing new. Freight elevators had existed since ancient times. Louis XV had even had a "flying chair" installed at Versailles, and steam-powered hoists, invented in England in the 1830s, were a common sight in America.

What Elisha Otis did that *was* new was to make it safe. He cut notches all along the wooden guide rails, or "shaft," of his hoist, then ran the hoist's ropes through a pair of springs he had screwed into the bottom of the elevator platform—a design probably inspired by the wagon brakes he had worked with years before. If the rope was cut, or broke, the spring opened and caught on the notches, stopping the elevator almost instantly.

There were already safety devices on some freight elevators, but these were usually rotating pinions, round gears that had to be worked by hand. Otis's safety brake did not depend on human reaction time.

He quickly sold three of his "automatics" to local merchants at $300 apiece. Then sales dried up. And his latest factory went belly up. Benjamin Newhouse, a furniture maker who'd bought Otis's first elevator, gave him a space at his Yonkers plant to build more. There Otis worked away with just some vises, a drill press, a forge, a used lathe, and a three-horsepower steam engine, distracted by the other inventions swirling in his head, doodling pious little maxims to himself. There were still no sales, and Otis at last seriously contemplated what Americans do when things go bad, which was to pull up stakes and head west.

He was detained by "the Prince of Humbugs" (God works in mysterious ways). P. T. Barnum needed a new attraction for the world's fair he had taken charge of in New York's grand Crystal Palace on Forty-Second Street. Otis, amazingly enough, proved a natural showman. Dressed formally in a burgundy topcoat with velvet lapels and his constant stovepipe hat, he would be hoisted high above the crowd on an open platform, while Barnum gravely advised those onlookers "prone to fainting" to "take out your [smelling] salts."

Then Otis would dramatically slash the rope of his elevator with an ax or a saber, drawing screams and cries. The brakes caught at once, every time. Otis would take off his hat, bow, and announce, "All safe, ladies and gentlemen, all safe."

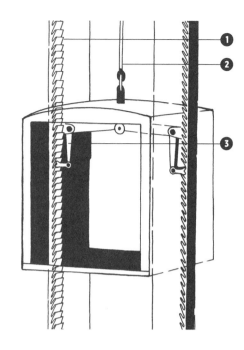

The Otis safety elevator featured notches (1) all along its hoist, and ropes (2) connected to a pair of springs (3) in the elevator platform. If the ropes tore or were cut, the springs would automatically open, catching the elevator on the hoist's notches.

After that, the orders poured in. Otis custom-built elevators for one store after another in Manhattan, including one with its own ingenious steam engine that enabled it to be almost instantly stopped or sent up or down. His grown sons chafed under his erratic direction—Charles was once forced to sign a statement pledging, "It is understood that I am not to volunteer advice or opinions concerning that part of the business not placed in my charge"—but once he was gone they would revolutionize the business, making Otis Elevator a household name around the world.

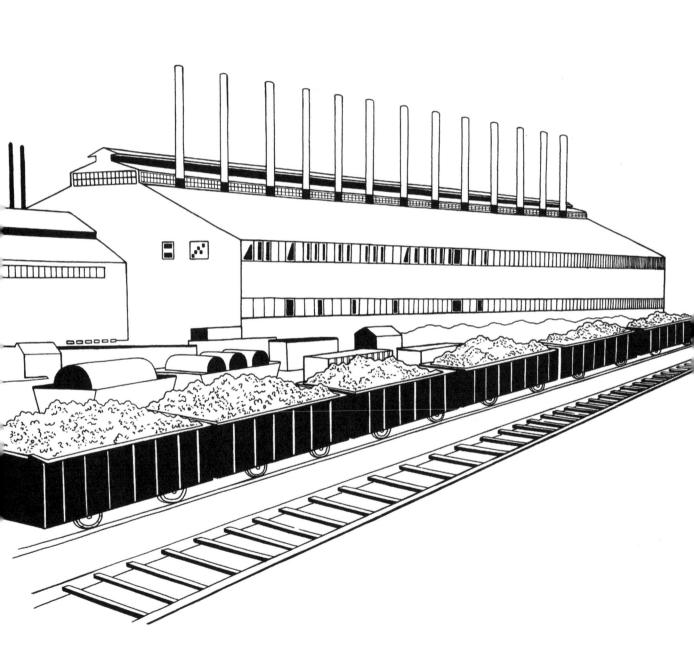

CARNEGIE'S STEEL MILLS

T he young man noticed things—things that almost no one else seemed to pay attention to, such as how often the iron rails on the Pittsburgh Division of the Pennsylvania Railroad cracked and caused derailments. They needed to be replaced every six weeks to two months, especially around curves and under the heavy industrial loads that freight trains carried.

Surely there had to be a better material that could be used—but what? There was steel, an intriguing new metal, but steelmaking was considered too laborious and expensive for anything but cutlery and specialized tools.

Andrew Carnegie tucked his observations away for a moment when he might do more about them. He had made a career of paying attention, ever since he'd come to the United States with his impoverished Scottish family in 1848. Put to work by age thirteen, he labored for pennies a day as a bobbin boy in a cotton mill, then as a telegraph messenger, but he had come to the right part of the world for an ambitious boy who noticed things: a western Pennsylvania that was brimming with invention and opportunity. Andrew memorized the faces of the city's most important men and worked constantly to "improve himself." His industry and desire impressed older, powerful men—and he made sure it did. By just eighteen he was in charge of the Pennsy's Pittsburgh operations, then superintendent of the Union's military railways and its eastern telegraph lines during the Civil War.

By the time he was twenty-five, Carnegie had gained a major holding in the Pullman Car company. Soon he was also invested in the new Pennsylvania oil fields, along with steel mills and iron works, a bridge manufacturing company, and iron ore fields around Lake Superior.

"I determined that the proper policy was 'to put all good eggs in one basket and then watch that basket,'" he would famously remark.

It was a risible misrepresentation—perhaps a purposeful one—of what he was really about. Like the other great American capitalists of his day, Carnegie kept his eye, always, on the connections between all eggs, and all baskets.

Carnegie's steel plant, the greatest American industrial plant of its time.

The nexus, for him, was steel. He had not forgotten how badly iron rails performed, and as soon as he had a chance Carnegie sponsored experiments in steel-coated rails, which proved exponentially more durable. But where to get enough of the metal?

Carnegie traveled to England in 1872 to examine the latest technology: the Bessemer Steel Converter. Looking to make massive cannons, the Sheffield metallurgist Henry Bessemer and his collaborator Robert Forester Mushet had developed a method of pouring molten pig iron into a converter shaped like a giant black egg and blasting it with air to remove the carbon impurities. Carbon and manganese were then added to the molten metal in exact amounts, according to a formula developed by Mushet. The whole process took just ten to twenty minutes to "cook" massive quantities of incredibly durable steel, and for as little as a sixth of the previous cost. Steel rails Mushet had first put down in 1857 had endured, despite seven hundred trains a day passing over them.

"I had not failed to notice the growth of the Bessemer process," Carnegie would later write. "If this proved successful I knew that iron was destined to give place to steel; that the Iron Age would pass away and the Steel Age take its place."

Carnegie returned to the United States to recruit partners and build a $1.2 million steel mill in the Pittsburgh community of North Braddock, along the banks of the Monongahela River. The workmen dug old bayonets and swords out of the earth. They were on Braddock's Field, where the French and Indians had ambushed the British General Braddock in 1755 and where George Washington had won the moniker of "Hero of the Monongahela." It seemed an auspicious place to start a new age, and Carnegie began to erect an industrial plant on a scale not really seen in America before—miles of gigantic furnaces and coke ovens, long tin sheds and conveyor belts, train yards and river wharves and towering smokestacks, rolling mills to convert the finished steel into rails. And of course the great black egg that would cook his steel.

Then the Wall Street Panic of 1873 sent American capitalism reeling. Carnegie was forced into a two-year battle to keep his new steel mill on schedule and in his own hands. Orders for rails and steel beams fizzled out as the economy melted down. Carnegie pursued his debtors relentlessly, staved off his creditors, and bailed two of his partners out of their obligations.

On August 22, 1875, he brought his great black egg hissing to life, turning out the first steel for an order of two thousand rails for the Pennsylvania Railroad. It was a dramatic process, one that produced "a ferocious geyser of saffron and sapphire flame," according to an 1893 description from *McClure's Magazine*, and created a "scene [that] became gorgeous beyond belief." It was also one that Americans would immediately identify with—making steel, mastering the elements.

It was a triumph of both corporate and technological design. On the business side, Carnegie had built himself a perfect vertical monopoly: Carnegie steamships brought Carnegie coal and iron across the Great Lakes from the Iron Range to be smelted into steel at Carnegie mills. He now had the leverage to outmaneuver and absorb his rivals, until by the 1880s Carnegie steel was producing more steel rails, pig iron, and coke than any other firm in the world.

The Bessemer process kicked open the door to modern industrialization, making steel roughly as cheap as wrought iron and infinitely preferable. Carnegie and his competitors were soon manufacturing countless miles of steel cable, used to build suspension bridges, such as the wondrous new structure connecting Brooklyn and Manhattan. Steel rods and sheet steel went into making swift new ocean liners; immense new boilers and locomotives; the turbines and generators used to convert the power of gigantic new dams; and pretty much everything else from the Ferris wheel to the locks of the Panama Canal. Steel beams made possible an entirely new sort of building (see page 135), one that could reach previously unheard-of heights.

By 1907, in the relentless competition of American industry, the Bessemer process had been surpassed in production by open-hearth steel, a slower but purer process that yielded steel with a higher tensile strength. Carnegie would sell his interests to the conglomerate that became the behemoth U.S. Steel, devoting the rest of his life to philanthropy. But by now America's spine was made of steel.

THE GENIUS DETAILS

A related process of removing carbon with air was used in steel production in China as early as the eleventh century and in Japan as early as the seventeenth century, but never on an industrial level.

No fewer than seventeen of the first "Chicago School" skyscrapers used steel from Carnegie's Freedom Iron Works.

The open-hearth process itself would eventually be largely replaced by the basic oxygen process, a steelmaking process similar to the Bessemer process. The Bessemer process became obsolete in the United States in the 1930s and was last used in America in 1968. A last Bessemer shop remains in the Russian Urals.

Carnegie sold his steel interests for $480 million, or over $300 billion in today's dollars, to financier J. P. Morgan in 1901. It was the largest personal sale in world history.

From 1875 to 1920, US steel production grew from 380,000 tons a year to 60 million. Production increased by an average of 7 percent a year from 1870 to 1913. By 1889, the United States was the largest steel producer in the world.

STEEL-FRAME CONSTRUCTION

T hey began to rise out of the prairie by Lake Michigan: buildings taller than anything but the highest cathedrals. Buildings that could be constructed upside down and that were just as thick on top as they were at the base. Buildings the likes of which nobody had ever seen before.

Chicago in the 1880s was the perfect place to build, mostly because the whole place had burned down in 1871 and had never really been rebuilt. Now there was a construction frenzy, spurred by an America that was about to become the world's leading economy and by the influx of immigrants pouring into the Midwest by rail.

"Chicago has thus far had but three directions, north, south and west," proclaimed the *Chicago Tribune*, "but there are indications now that a fourth is to be added . . . zenithward."

"Zenithward" it would be. In 1882, a pair of immortals in what would become known as the "Chicago School" of architecture, Daniel Burnham and John Wellborn Root Jr., erected the ten-story Montauk, the first building to be called a "skyscraper." It made revolutionary use of fireproofing materials and solved one of the problems of building tall in Chicago's soggy, shifting soil when Root invented an ingenious "floating raft"—a foundation of steel rails wrapped in concrete, upon which the building could distribute its weight evenly.

Yet the Montauk had essentially been built as every other building in history had been, with load-bearing masonry walls. The limitations of this for going zenithward became clear in another Burnham and Root creation, the sixteen-story Monadnock Building, which had ground-floor walls that were six feet thick and took up 15 percent of the floor area. Wrought-iron frames were one alternative, but iron was massive, not terribly flexible, and could become brittle and collapse without warning.

The answer was provided by the man who had taught Burnham—as well as Louis Sullivan, and many other key architects in the Chicago School. Major William Le Baron Jenney was an eccentric figure, affectionately described by Sullivan as "monstrously pop-eyed, with hanging mobile features, sensuous lips," someone whose "English speech jerked about as if it had St. Vitus's dance."

Chicago's New York Life Building, the first true skyscraper, and today a four-star hotel.

Jenney had been born just outside New Bedford, Massachusetts, to a father who owned a fleet of whaling ships (see page 159), and thereafter his life had been one long adventure. He had mined for gold in California, engineered a rail line across lower Mexico, and sailed on one of his father's whalers to the Philippines, where he saw treehouse constructions that might have helped inspire his skyscraper designs.

In the course of his fantastic career, "the Major" would design parks, private homes, and office buildings, his designs and creations anticipating the work of Frank Lloyd Wright's "Prairie School," the Bauhaus, and even the "glass curtain" style of modern architecture. Erecting a ten-story branch-office tower in Chicago for the Home Insurance Company of New York in 1884, Jenney built the first six stories out of iron from Andrew Carnegie—then finished the top four stories with Carnegie's Bessemer steel beams (see page 131).

This, agree most architectural critics, was the first real skyscraper—at 138 feet high, the

Jenney's New York Life Building under construction, with the granite and terra-cotta cladding added in no particular order to the steel frame, which holds it all up.

tallest building in the world. More important, the structure was almost entirely supported by its inner frame. (Part of it is thought to have rested on granite piers and a rear brick wall.) Weighing only about a third of what a comparable stone building would have, the Home Insurance Building also featured fireproofing, modern plumbing, safety elevators (see page 127), and wind bracing. Still, its radical method of construction so worried Chicago officials that they stopped construction for a time to thoroughly inspect it.

"For the first time since bricks were burned in history's early dawn," Jenney's contractor, Henry Ericsson, would write in his autobiography, "men were laying brick walls that were only curtains. . . . A metal frame now supported both floors and building load."

The Major was just getting started. Over his next several commissions, he pushed his buildings still higher, inventing and perfecting the "birdcage" style of steel-frame architecture, in which the lintels were extended over the windows to connect with the steel columns and floor beams. Thanks to the rapid production, strength, and flexibility of steel, Jenney could build with tremendous speed and power, putting up the Fair Store, an eleven-story department store, in just three and a half months.

With the New York Life Building, though, Jenney outdid himself. Finishing its twelve-story birdcage skeleton before all the gray Maine granite that was to cover its first three floors was ready, Jenney began to add the glazed terra-cotta "skin" planned for the upper floors, almost willy-nilly. It no longer mattered when you put the outside cover of the building up; that was just for ornamentation and human comfort. It had nothing to do with keeping the thing up. No building had ever been built like this, and it ushered in what Jenney called "an age of steel and clay," with the light, malleable terra-cotta molded into all sorts of classical ornamentation.

The world's leading architects and engineers were streaming into Chicago at the time to see the Columbian Exposition's "White City"—the gleaming Beaux-Arts neo-classical idealization of what a city might be, much of it supervised by Jenney and some of it built by him. On the way, they got to gape at his New York Life Building, being finished from the top down as well as the bottom up.

In the years ahead, Chicago would lose its lead in skyscrapers and some of its greatest architects to New York, with its harder bedrock and more restricted space. New York authorities distrusted Jenney's "birdcage" construction, mandating "cage construction" instead, in which a birdcage structure supported the interior but the steel exterior walls supported everything else.

This enabled construction of the tallest buildings ever erected: the Metropolitan Life Insurance Company Building, nearly seven hundred feet high, with clock faces bigger than that on Big Ben and a searchlight that would signal weather conditions to ships far out at sea; Cass Gilbert's sixty-story Woolworth Building, "the Cathedral of Commerce," with its gorgeous Beaux-Arts interiors and terra-cotta craftsmen placed like saints at its entrance; Daniel Burnham's fabulous "Flatiron" Building, plowing into Madison Square, in the words of its greatest portraitist, Alfred Stieglitz, "like the bow of a monster ocean steamer."

The triangular shape of the Flatiron fascinated and appalled New Yorkers, leaving some of them too scared to walk close to the thing, especially when its cladding was sometimes applied, like the skin of the New York Life Building, from the top down. For residents of Jenney's Chicago, this miracle was old hat.

Architectural historian Carl Condit called William Jenney's 1884 Home Insurance Building the most important innovation in architecture since the introduction of the Gothic cathedral in the twelfth century.

Completed in 1894, the New York Life Building survives to this day on the corner of LaSalle and Monroe Streets in Chicago.

By 1903, the New York Life Building had been raised from twelve to fourteen stories, and its frontage expanded from 141 to 233 feet.

The New York Life Building's "elegant lobby is notable for its superb design and craftsmanship in sumptuous materials." Two stories high, the lobby was built with gray marble–clad walls and a coffered ceiling.

Some maintain that the beams of Jenney's Home Insurance Building—demolished in 1931—were not wholly self-supporting and that thus Burnham and Root's ten-story, all-steel construction, the Rand McNally Building (1889–1911), should be considered the first true skyscraper.

COOLING THE AIR

P eople have been trying to cool the air since at least the days of the pharaohs, when ancient Egyptians hung reeds soaked in water in the windows of their mud huts. For a very long time, most efforts at changing indoor air temperature everywhere—from Rome to China, Paris to Persia—were concentrated along the same basic line: running air over water.

A more sophisticated method of cooling air was demonstrated in 1758 when Benjamin Franklin and a Cambridge chemistry professor named John Hadley— building on the work of the Scottish scholar William Cullen—proved that the rapid evaporation of volatile liquids, such as alcohol or ether, would absorb heat from surrounding air or objects, thereby cooling them. Scientists in America, in England, and on the European continent soon grasped the potential this offered for refrigeration, but it wasn't until 1834 that the versatile American inventor Jacob Perkins was able to invent a closed-loop, vapor-compression refrigeration system in which liquid was continuously evaporated into gas, then changed back again to liquid, cooling the air around it.

Just a few years later, Dr. John Gorrie, resident physician in the Florida cotton town of Apalachicola, was desperately trying to bring relief to patients suffering from yellow fever and malaria in his hospital. Working obsessively, Gorrie spent years building a machine that would make ice by compressing air, then expanding it around surrounding water tubes. His ice machine was much like Perkins's invention, but the doctor added a new component: a fan that would blow the cooled air over the ice and thus make bearable even a Florida hospital room in the summer. Unfortunately, his invention, too, went nowhere, and Gorrie died a few years later, shattered financially and emotionally.

As we all know, it's not so much the heat as the humidity. Nearly a half century after Gorrie's death, Brooklyn's Sackett-Wilhelms Lithographing & Publishing Company was complaining that the region's legendary humidity was ruining its meticulous multicolor prints by warping the paper and causing the inks to misalign. The problem was outsourced to Willis Haviland Carrier, a young man with a master's in mechanical engineering, who was then making all of ten dollars a week for the Buffalo Forge Company.

They had gone to the right man. It had recently occurred to Carrier, while standing on a foggy Pittsburgh train platform, that by passing dry air through water he could *create* fog—that is to say, air with a specific amount of moisture in it. What he translated this into was an air-conditioning system built on much the same principles Perkins and Gorrie had used. An air compressor cycled refrigerant continuously from an evaporator through a condenser and an expansion valve and then back to the evaporator. In so doing, the refrigerant changed from a liquid to a gas, then back again. At the same time, a fan pulled hot air from a room, and the evaporator pulled the heat from the air—just as Ben Franklin discovered—and condensed its moisture into water. Fans then sent the cool, dry air back into the room and expelled the heated air through a vent to the outside.

Carrier was able to drop the humidity at Sackett-Wilhelms to just 55 percent. Within a decade, he came up with his "Rational Psychrometric Formula," calculations for keeping humidity steady that are used in air-conditioning to this day, and soon he was designing air-conditioning systems for all sorts of factories, bakeries, and food-processing plants. Leaving Buffalo Forge to build the industrial giant that still bears his name, Carrier invented a centrifugal chiller that enabled his company to air-condition enormous indoor spaces: Macy's, and the Hudson Department Store in Detroit; the US Congress and the White House; even Madison Square Garden.

Many challenges still remained before air-conditioning could become a routine part of American life. Early coolants, such as ammonia, ethyl chloride, sulfur dioxide, and propane, could be highly flammable, toxic, or even lethal when they leaked. These were replaced by much less dangerous chemicals such as dielene and DuPont's Freon (which would later prove to have its own very big problem as a chlorofluorocarbon, endangering the earth's vital ozone layer).

Early industrial air-conditioning units were also enormous. They would be steadily reduced in size to

Carrier built his air conditioner on his first work assignment after graduating from college.

Carrier's first air conditioner for the Sackett-Wilhelms plant provided the cooling equivalent of melting 108,000 pounds of ice a day.

The percentage of Americans living in the southern and southwestern states increased from 28 percent of the total population in 1950 to 40 percent by 2000.

The first home air-conditioning unit was commissioned by Minneapolis millionaire Charles Gates, son of the famous gambler and stock plunger John "Bet-a-Million" Gates. It was seven feet high, six feet wide, and twenty feet long, and may never have been used, as the house it was designed to cool was never lived in.

Stuart Cramer invented the term *air-conditioning* in 1906 for his system of cooling a North Carolina textile mill by introducing cooled water into the air.

make them practical for cooling movie theaters, offices, apartment buildings, buses, and railroad cars, but it wasn't until 1932 that your basic window-ledge AC unit came on the market—available for an astronomical $10,000 to $50,000. Only at the end of World War II was the first affordable portable air conditioner invented.

The progress of air-conditioning would prove unstoppable, though, and it would transform the country. Without air-conditioning, the enormous postwar and population booms throughout the South and Southwest would have been inconceivable. Entire cities, such as Las Vegas and Phoenix, would be able to rise only with the advent of universal air-conditioning, while AC would also change the tenor of long, hot summers even in more temperate regions, ending such traditions as sleeping on roofs or fire escapes. Our whole culture would shift to one more oriented to the great indoors, where it was always a comfortable sixty-eight degrees. Air-conditioning would be so indelibly associated with America, so long before the rest of the world, that it would come to seem the quintessentially audacious American idea. To contradict one of our greatest writers, if you don't like the weather, you *can* do something about it.

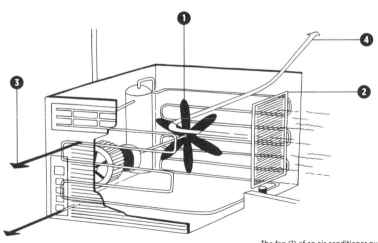

The fan (1) of an air conditioner pulls hot air from a room, and the evaporator (2) then pulls the heat from the air by continuously converting a refrigerant from a liquid to a gas and back again. The cooled air is fanned back into the room (3), while the heat is discharged to the outside (4).

THE GOLDEN GATE BRIDGE

There is no more beautiful or dramatic American invention than the steel-wire suspension bridge. Fantastically simple—in theory, anyway—suspension bridges go back at least to 1433, when the Buddhist saint Thangtong Gyalpo created iron-chain bridges to span the chasms of Bhutan. For the next 450 years, though, they remained mostly modest foot- or wagon bridges until a German immigrant, John Augustus Roebling, and his son, Washington, spanned New York's East River.

The Brooklyn Bridge was a seminal achievement, a brutally difficult undertaking that took fourteen years, sparked a revolution in building, and eventually turned New York into a city of bridges, with sleek, bejeweled spans adorning it like so many diamond necklaces.

Yet no location better frames the suspension bridge than the Golden Gate of San Francisco Bay, portal to the Pacific. Its realization is the story of how nearly every element of a single American community—a team of unconventional engineers, a legendary banker, a near madman in charge, and the people themselves, willing to risk their own homes and businesses—came together to bring about something of surpassing beauty and utility.

Fitting a bridge across the three-mile Golden Gate Strait faced many seemingly insurmountable obstacles. The navy didn't want it, fearing any bridge would interfere with navigation. The powerful Southern Pacific Railroad didn't want it, fearing it would hurt its ferry fleet. The Sierra Club was afraid of the environmental damage. Politicians balked at the cost, calculated at $100 million (over $2 billion today), while engineers pointed out that it would have to be the longest suspension bridge ever constructed, at 4,200 feet, and the first built with its towers exposed to the open sea.

The bridge's only allies were a few newspapers, the still fledgling automobile industry, and the people of Northern California. It was their pressure that got the state legislature to create a Golden Gate Bridge District in 1923, the board of which hired Joseph P. Strauss to be chief engineer.

It was, to say the least, an unconventional choice. Strauss was short on the experience needed, high on determination. He lied to the board about having a graduate certificate in engineering. A frenetic, hyperbolic individual from Cincinnati who

The Golden Gate Bridge, stretching 1.7 miles from San Francisco to the Marin Headlands.

loved to write poetry, he seemed not quite in his right mind much of the time. Just five-three, he insisted on going out for the University of Cincinnati's football team. When this landed him in the hospital, he fell in love with a Roebling suspension bridge across the Ohio that he could see from his window. For his senior thesis, he presented a scheme for a fifty-five-mile rail suspension bridge across the Bering Strait, bemusing the gathered faculty and fellow students but leaving the hall ringing with enthusiastic applause.

Love suspension bridges though he might, Strauss spent his whole career designing and building hundreds of bascule bridges (drawbridges) in the Midwest. His first idea for the Golden Gate Bridge was another such structure, ugly and awkward.

Fortunately, he had assembled his own unconventional team of engineers and architects to support him. The final design was drawn up by Leon Moisseiff, the Jewish Latvian immigrant who had designed the Manhattan Bridge back in 1910. Checking the structural engineering was the Swiss immigrant Othmar Ammann, who had just designed the stunning George Washington Bridge over the Hudson and who would eventually ring New York with six such spans, including the Verrazano-Narrows. Irving Morrow, previously an apartment house architect, provided the art deco tower decorations, streetlights, railings, and walkways and the distinctive "International Orange"—that is to say, "red"—paint color that make the Golden Gate a sculpture as much as a bridge.

Principal engineer on the project was Charles Alton Ellis, a scholar in ancient Greek and mathematics, who had written the standard textbook on structural engineering and become an engineering professor—before finishing his engineering degree. Ellis compiled ten and a half volumes of higher mathematical calculations with just a circular slide rule and a hand-cranked adding machine, to clarify the thirty-five separate "unknown units" (factors) about the bridge's viability. In his megalomania, Strauss fired him from the project and tried to eradicate all record of his contributions. Badly hurt, Ellis became convinced that the bridge contained fatal design flaws.

He was wrong, thanks to the brilliance of Moisseiff, whose "deflection theory" allowed the Golden Gate Bridge's roadway to sway nearly twenty-eight feet in either direction in the wind, thereby transferring stress to the steel suspension cables—spun by the Roebling Company—that could easily handle it.

Gorgeous it was, safe it was, but as planning ran smack into the Depression the question became who was going to pay for it. Strauss had wrangled the cost down to $35 million. The people of the district had taxed themselves to raise the preliminary work and had raised the money to issue bonds, in some places, by offering their own homes and farms as collateral. But by 1932, with the economy crumbling, no bank could be found to float the bonds. Saving the day was the founder of Bank of America, Amadeo Peter Giannini, a volatile, steel-willed former fruit salesman, child of Italian immigrants to San Francisco, who had done much to build modern California. Giannini had his bank take on the entire issue of bonds just to provide work to the town where he had made his fortune.

When Giannini asked how long the bridge would last, Strauss told him, "Forever."

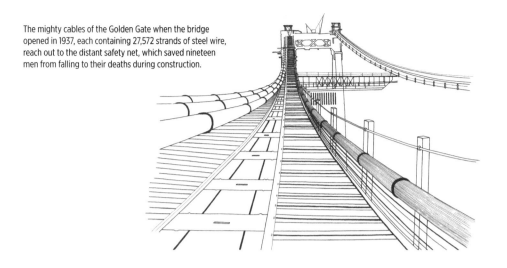

The mighty cables of the Golden Gate when the bridge opened in 1937, each containing 27,572 strands of steel wire, reach out to the distant safety net, which saved nineteen men from falling to their deaths during construction.

The work of spanning the Golden Gate was dangerous in the extreme, thanks to winds that could reach 60 miles an hour, fog that turned the work surfaces to ice, and the churning deep water below. For all his egomania, Strauss cared enough about the workers building his bridge to put a revolutionary new system of safety netting below them. Even this failed, near the end of construction, when the netting's bolts popped after a scaffold collapsed, hurtling twelve men and a platform into the net at once. Ten died, but thanks to the net breaking their fall, two managed to survive the 200-foot plunge into the icy, fast-moving currents below. Over the course of the work, another nineteen men were saved by the net, forming what they dubbed the "Halfway to Hell Club."

An incredible amount of work had to be done before the project itself could even begin. Building the south (San Francisco side) tower required constructing first a protective, concrete "fender," 300 feet long, 155 feet wide, and consisting of 152,600 tons of concrete—a thousand feet out into open sea, which was 100 feet deep. Both bridge towers had to be bored into a solid rock ridge 235 feet below the surface of the ocean, while the bridge's roadway was suspended from two main cables, which had to be passed through the towers and fixed in concrete on each side of the strait.

On May 27, 1937, the Golden Gate Bridge was finished—$1.3 million *under* budget. Two hundred thousand people walked across it, gazing in wonder at what they had wrought. The *San Francisco Chronicle* called it "a thirty-five million dollar steel harp!"

By the time the bridge was finished, Strauss had a little less than a year to live. But he did see the poem he'd written to celebrate the occasion, "The Mighty Task Is Done," fastened onto one of the bridge towers:

> ASK OF THE MIND, THE HAND, THE HEART,
> ASK OF EACH SINGLE, STALWART PART,
> WHAT GAVE IT FORCE AND POWER . . .

UTILITY AS BEAUTY
THE RIVER ROUGE

By 1915, the great man was restless again. All his life, Henry Ford would move abruptly, sometimes irrationally, from one obsession to another—but always with the dream of building an empire beyond most other men's imagination.

Ford had created the greatest car company in the world almost by hand at times, assembling, racing, and even advertising his own automobiles. His rough black beast, the Model T, had transformed American life—the first affordable, gas-fueled car. Contrary to popular belief, Ford did not invent the assembly line, but he had taken Eli Whitney's and others' (see page 205) "American system" of mass production through interchangeable parts to new heights—then made it possible for countless Ford workers to buy his cars by doubling the standard industrial wage.

Now he moved to acquire two thousand acres of bottomland where the Detroit River met the Rouge, in Dearborn, Michigan—an area so unprepossessing it was thought he intended to establish a bird sanctuary there. Instead, it was the first step in realizing another vision: an "ore to assembly" auto plant that would, as much as possible, manufacture his cars where the raw materials to make them came in, then ship them from there right out to the market. What he had in mind became known as "Fordism": the vertical control and integration of all the raw materials, all the manufacturing processes, and all the men in the thousand different jobs needed to conceive, create, and assemble every car from scratch.

After dredging the Rouge River, Henry began to build docks for a fleet of Ford ships, bringing in lumber from seven hundred thousand acres of Ford-owned forests in the Upper Peninsula of Michigan and iron ore and limestone from Ford quarries in Minnesota and Wisconsin. He erected depots for Ford freight trains, bringing in coal from Ford mines in Kentucky, West Virginia, and Pennsylvania. A brilliant architect, Albert Kahn, "the man who built Detroit," was hired to design dozens of factories, mills, and plants that were considered state-of-the-art workplaces, full of light and air. With all the room available, Ford could shift from the multistory means of production at his old Highland Park factory to more efficient single-story branch assembly, building everything faster, cheaper, better.

The most complete production and assembly plant ever built in America.

Ford used the very first Rouge plants to build not cars but Eagle antisubmarine boats for World War I, in a deal he struck with an ambitious undersecretary of the navy named Franklin Delano Roosevelt. Next he churned out hundreds of thousands of mechanized "Fordson" tractors, sold primarily in Europe and the Soviet Union. While the Rouge made all the parts for the Model T, cars were still not assembled at the new plant until the "Tin Lizzie" was finally retired in 1927 and Ford rolled out its new Model A, followed soon by the world's first true "power cars," with their model V-8 engines.

Nearly everything was done onsite. Iron smelted in the Rouge's coking ovens and immense blast furnaces went straight to the largest foundry on earth, where it was poured into molds for engine blocks, cylinder heads, and other auto parts—close enough so it did not have to be reheated. Steel was forged, rolled, and stamped in the Rouge's mills. More plants made tires; built transmissions, radiators, and batteries; created tools and dies; produced paper; carved wood; converted soybeans into plastic; wove textiles into upholstery; and even made the world's first auto "safety glass." A power plant produced as much electricity as was needed to light Detroit at the time—a city of over a million people.

By 1928, when it was completed, the Rouge was far and away the largest integrated factory in the history of the world and "without parallel in sheer mechanical efficiency," according to historian David L. Lewis. It contained a dizzying array of 93 buildings, 120 miles of crisscrossing conveyor belts, and nearly 16 million feet of factory floor space, an industrial plant so awe-inspiring in its visible power and beautiful in its design that it became something else as well: art.

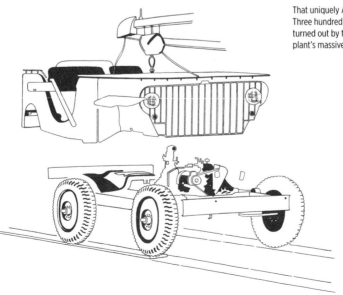

That uniquely American vehicle, the Jeep. Three hundred thousand of them were turned out by the Rouge as part of the plant's massive contribution to the war effort.

Thanks to Henry Ford's much-harried son, Edsel, the River Rouge would be immortalized by the American artist Charles Sheeler in a series of stunning, black-and-white photographs and vividly colored paintings, and then—most incongruously—by the great Mexican muralist Diego Rivera, a self-proclaimed communist.

Touring the industrial creations of the United States, Rivera became convinced that "in all the constructions of man's past . . . there is nothing to equal these." He told a reporter, "Here it is—the might, the power, the energy, the sadness, the glory, the youthfulness of our lands." What he produced in its honor would be one of the great creations of American art, his fresco of the Rouge, on the walls of the Detroit Institute of Art.

The River Rouge would be the last great creation of Henry Ford. His once unfettered mind was mired in a swamp of paranoia, bigotry, and perhaps dementia. Fordism—especially the ceaseless demand of the assembly line—was always hard on Ford workers, and their boss would hire thugs and mobsters to suppress any dissent and spy into every corner of their lives.

Yet the River Rouge would also become an icon of workplace democracy, where United Auto Workers Union (UAW) leader Walter Reuther survived a brutal beating to win the right for Ford workers to unionize. Reuther's enlightened form of unionism—as forward thinking as Ford's had been in production—would help bring the company back from the brink of collapse. The River Rouge would help the United States win World War II and make Ford an automotive competitor again after the war, turning out such now classic cars as the Thunderbird and the Mustang.

Today, there are still six thousand employees at what is now the Ford Rouge Center, most of them making trucks. It is a work environment emphasizing environmental sustainability, "green" technology, decentralization, teamwork, a wide array of suppliers, and "just-in-time production"—almost the exact opposite ideals of the beautiful behemoth Henry Ford built, but a plant for the new century.

THE GENIUS DETAILS

The Rouge was a city of one hundred thousand workers at its peak in the 1930s, able to turn out a new car every forty-nine seconds.

In the course of a good year, Ford used five hundred thousand freight cars to ship materials and parts and ran up $150 million in shipping costs.

The Rouge had its own internal railroad, with one hundred miles of track and sixteen locomotives; its own bus system for moving people around fifteen miles of paved roads; and its own fire department, with multiple stations, police department, hospital, and maintenance crew of five thousand.

During World War II, the River Rouge plant turned out amphibious vehicles, tanks and tank engines, aircraft engines and nearly three hundred thousand Jeeps.

The current Ford Rouge Center consists of only six hundred acres, or less than a third of the old River Rouge plant, but it is still the company's largest facility.

"ARCHITECTS MAY COME,
AND ARCHITECTS MAY GO . . ."
FALLINGWATER

B y the mid-1930s, Frank Lloyd Wright's star seemed permanently dimmed. Proclaimed as the boldest young builder in America twenty-five years before, thanks to his leading role in establishing the "Prairie School" style of architecture, he had been lashed by a series of punishing setbacks ever since. His lover, her children, and four other people had been murdered at his famed estate and studio in Taliesin, Wisconsin, by a deranged servant, who then burned much of the place to the ground. Wright rebuilt it—only to see most of it burn again in a freak electrical fire. The bank almost foreclosed on a third version as the Great Depression and public umbrage at Wright's scandalous personal life dried up commissions and forced him to sell much of his priceless collection of Japanese art.

Teetering on bankruptcy, nearing seventy years of age, Wright was approached by a former apprentice of his, Edgar Kaufmann Jr., to build a new rural retreat for his father, owner of the biggest department store in Pittsburgh. What Edgar Sr. had in mind was a modest family vacation house at the bottom of a small waterfall on Bear Run Creek, where the family loved to cavort in the water and sun on the rocks.

What Wright gave them instead was a masterpiece, what has been called "the best-known private home for someone not of royal blood in the history of the world." Working in his usual bravura style, Wright visited the area, ordered up a topographical survey, then let it all gestate for the next nine months. Out of the blue, Kaufmann called him while on a trip to Milwaukee and said he was eager to see the plans. Wright, who had committed nothing to paper thus far, calmly told him to come ahead. While his apprentices looked on in astonishment, he drew the plans for what would become Fallingwater in the two hours it took Kaufmann to drive out to Taliesin, putting his signature on the drawings just as the department store magnate walked into the room.

Kaufmann was not pleased. Instead of building a house to face the falls, Wright had designed Fallingwater to sit directly *over* them, its decks anchored into the cliffside with a series of cantilevered concrete supports and steel beams. A rocky

Fallingwater, "the best-known private house for someone not of royal blood in the history of the world," according to author William Allin Storrer.

construction process unfolded in the next three years as Kaufmann and Wright battled over how thick the supports should be. Wright's works were never noted for their flawlessness, and Fallingwater's propensity to leak and collect mold led Kaufmann to later dub it "a seven-bucket building" and "Rising Mildew." By the 1990s, the house's steel and concrete reinforcements required a major overhaul to keep it from literally falling apart. Commissioned for $35,000, Fallingwater would end up costing Kaufmann $155,000, or about $2.6 million today.

What he had bought was the work of America's most creative architect at the height of his power. Fallingwater represented all that Wright had learned from studying and building for decades, both on the American prairie and in Japan. Its only man-made colors are Wright's trademark, earthy "Cherokee red" steel supports and the ocher concrete, the shade of a fallen rhododendron leaf. Angled to bring the sun deep into all rooms during the day, the glass of its columns and bands of windows seems to melt away at night. At all times, the house is filled with the sound of running water. While its decks and balconies jut out into the woods, an open staircase leads directly to the

The color of Fallingwater, the materials used in its construction, and its location in the midst of a waterfall all make it appear to fit seamlessly into the natural world around it.

creek below, and Edgar Sr.'s favorite sunning rock is actually incorporated *into* the living room hearth. It is an unsurpassed use of space, "an architecture [that] not only work[s] with nature . . . [but] almost become[s] part of nature," as Edgar Kaufmann Jr. put it.

Combined with the completions of his first "Usonian" houses and the magnificent Art Moderne Johnson Wax Building at about the same time, Fallingwater shot Frank Lloyd Wright back to the forefront of American architecture. He would remain there for the rest of his long life, to this day the most renowned builder in our history.

THE GENIUS DETAILS

Frank Lloyd Wright's fee for designing Fallingwater was $8,000.

Located forty-three miles from Pittsburgh in a remote corner of southwestern Pennsylvania, Fallingwater has nonetheless attracted over five million visitors since it opened to the public in 1964.

Wright never graduated from high school or college and never joined the American Institute of Architects (AIA), which he despised. In 1991 the AIA nonetheless voted Fallingwater "the best all-time work of American architecture."

Wright wrote of himself: "Not only do I fully intend to be the greatest architect who has yet lived, but fully intend to be the greatest architect who will ever live. Yes, I intend to be the greatest architect of all time."

Fallingwater inspired the fictional home of Phillip Vandamm, the villainous spy for the Soviets played by James Mason in Alfred Hitchcock's *North by Northwest*.

BUILDING THE BIG ONION FROM SCRATCH

W illiam Ogden was appalled. "[You have] been guilty of an act of great folly in making [this] purchase," Ogden chastised his brother-in-law, Charles Butler, in 1835, after he first laid eyes on the $100,000 worth of land Butler had bought just north of the Chicago River.

It was hard to blame him. When Ogden went out to assess his brother-in-law's holdings, he found himself in a marsh full of wild grass and sank knee-deep into the mud.

Water on the surrounding prairie was routinely three feet deep—and none was fit to drink. It had to be brought in by wagon for the "inmates," as Chicagoans liked to call themselves. Nothing grew there, save for the wild grass that kept all the mud from being washed into Lake Michigan. Food was hauled up from the Wabash Valley, two hundred miles to the south, but the wagons got sucked down into the same mud that accosted Ogden. In the winter, when temperatures reached twenty-eight degrees below zero, starving wolves prowled the outskirts of the town, searching for garbage, of which there was plenty.

Men had seen the commercial potential of Chicago's central, lakefront location ever since the French voyageur Louis Joliet reached it in 1673. But Chicago lacked any natural harbor, and a sandbar blocking the Chicago River made the unloading of goods or people difficult and dangerous.

More than any other great city in America, Chicago would be created by the works of men. And more than any other city, that creation would be the work of *one* man.

William Ogden would become Chicago's first mayor. He would also draft its city charter, finance its first canal and first successful railroad, run its first bank, design its first bridge, start its first steamship service, fund and head the board of its first university, donate the land for its foremost medical facility and its largest cathedral, recruit and sponsor its first great business, and build its first great house. He would be

Modern downtown Chicago, a great city built almost overnight from a desolate outpost. The flow of the Chicago River, which runs through it, was reversed by engineers in the 1880s to run from Lake Michigan to the Mississippi watershed.

An avid lobbyist for the Transcontinental Railroad, Ogden was the first president of the Union Pacific Railroad, and the "Golden Spike" that completed it was hammered through next to Ogden Flats, Utah.

Ogden hired a young lawyer named Abraham Lincoln to help him clear title to much of Chicago's waterfront.

With his mother and sister, Ogden threw lavish parties at his Greek Revival mansion for distinguished visitors including Daniel Webster, Ralph Waldo Emerson, Martin Van Buren, Samuel Tilden, Margaret Fuller, and William Cullen Bryant. "The guest always found good books, good pictures, good music, and the most kind and genial reception," recorded a neighbor, of Ogden's soirees.

Ogden organized the very first Chicago convention, consisting of over ten thousand investors and businessmen he gathered to push for more funds for the city's harbor and railroads. Chicago would become the meeting place of the nation, thanks to its rail connections, hosting countless conventions of every sort.

the city's most tireless booster, its most farseeing champion, the endower of his own, indomitable spirit.

"I was born close to a sawmill, was cradled in a sugar trough, christened in a mill pond, early left an orphan, graduated from a log schoolhouse and, at 14, found I could do anything I turned my hand to and that nothing was impossible," he wrote of his beginnings.

Born in upstate New York in 1805, Ogden took over his family's lumber and development businesses at sixteen. By twenty-nine, he was a successful businessman and well-connected member of the New York State Senate, where he lobbied hard for a railroad to the Great Lakes to augment De Witt Clinton's new Erie Canal (see page 9).

Once he'd knocked the mud from his boots—and sold just one-third of his brother-in-law's land for its entire purchase price—Ogden understood the potential Chicago had to extend what Clinton had done for New York City across the continent, making it the next hub in a global network of previously unimaginable wealth and commerce. On the recommendation of a young army engineer named Jefferson Davis, the federal government, in 1834, had cut a channel through the sandbar blocking the Chicago River and had lined the opening with piers, which was all Ogden needed.

As mayor and then alderman, Ogden paved roads, built sidewalks, designed what became the first swing bridge over the Chicago River, helped put together the Chicago Hydraulic Company to pump in water from Lake Michigan—even founded the town's first brewery for those still wary of the water. Soon ships were sailing in from the East, courtesy of the Chicago and Michigan Steam Boat Company, which Ogden organized.

Missing still was the vital link from Chicago to the rich farm and timberland of the surrounding Midwest. Ogden won support from the state to build the Illinois and Michigan Canal, a ninety-six-mile passage that would connect the city to the Mississippi, with as much potential for transforming the new city's fortunes as the Erie Canal had done for New York. Then a financial panic shut down any work on the canal, and Chicago's leading citizens wanted to repudiate the city's debts. Realizing this would destroy the town's future credit, Ogden backed the city's paper with his own money.

By 1848, the canal was finally finished. Ogden was already reorganizing a moribund railroad into the Galena and Chicago Union, going town to town along its proposed route, getting farmers' wives to use their egg-and-butter money to buy single shares. When the Chicago City Council still refused to let it in the city, Ogden got permission to bring a small rail engine, "the Pioneer," in from the harbor along "temporary rails." He then put on an instant demonstration of what the railroad would mean, running his engine eight miles out of town, borrowing a load of wheat from a farmer, and running it back into the city. Within a week, the ban on the railroad was lifted.

Chicago would soon become the rail hub of the entire nation. The question was what all these thousands of miles of rail were going to carry. Ogden brought in fifty million feet of timber a year to build the new city from two hundred thousand acres of pineland he had acquired along the Peshtigo River in Wisconsin.

Next was wheat. Ogden put up $50,000 to get Cyrus McCormick to locate his mechanical reaper company in Chicago (see page 208). Soon thousands of proud red McCormick's reapers were moving out across the prairie on Ogden's trains—and sending back over fifty million bushels of wheat by 1861, making Chicago the greatest wheat port in the world. The city that just thirty years before could not feed itself now had five years' worth of grain stored along its waterfront in twelve huge silos provided by yet another Ogden enterprise.

No great city ever grew so rapidly. The population of Chicago rose from about 200 in 1833 to over 112,000 in 1860, when it hosted the Republican convention that nominated Abraham Lincoln for president.

When the great fire of 1871 burned down the town he had done so much to build— and his timber holdings up in Peshtigo—Ogden, by then living in semiretirement in the Bronx, rushed west to see what he could do. What he found was little more than he had started with over thirty-five years earlier. His response was to help the survivors in both Peshtigo and Chicago and then to lobby the Illinois legislature in Springfield for money to rebuild.

Ogden's first draft of Chicago was erased. But his city was an irreplaceable linchpin of the greatest economy in the world at the height of the industrial age. It would shake off its terrible tragedy with remarkable ease, reaching a peak in population of 3.6 million by 1950, becoming again a Goliath of transportation and commerce, attracting immigrants and migrants from everywhere—even giving birth to a revolutionary architecture of remarkable grace and skill (see page 135). Perhaps Ogden's greatest gift was to endow his creation with the ability to do for itself.

THE WHALING SHIP

Men have been hunting whales almost since the first ship put to sea. It would take Yankee ingenuity and daring to convert their occupation into big business.

Whales in colonial days were so abundant in the North Atlantic that it was not uncommon for them to wash up on New England beaches. By 1793, though, the enormous animals had become wary of whaling ships, and when the New Bedford whaler *Rebecca* pursued its prey 'round the Horn that year, whaling was transformed from a fishing expedition to an epic adventure. Expeditions lasted three or four years, sailing to the Sea of Japan or the West Indian Ocean, and might include wintering in the Arctic. (One ship, the *Nile*, was at sea *eleven* years before returning to its home port in New London, Connecticut, though that may have been just what the whalers told their wives.)

By the golden age of American whaling in the 1850s, it was the fifth-largest industry in the United States, a $70 million business employing more than seventy thousand people. Of the estimated 900 whaling ships worldwide, 735 came from America—from little river towns off the Hudson and the Connecticut, from Long Island, and from the tiny atoll known as Nantucket. Four hundred ships sailed from the port of New Bedford, which called itself "the City That Lit the World" and was for a time the richest place, per capita, in America.

US dominance of the industry flowed from the boldness of its capitalists, often syndicates of parsimonious Quakers, and the innovations and almost lunatic bravery of its men. The basic techniques of harpooning and rendering whales had been established centuries before by Basque and Dutch whaling fleets, but Americans dramatically improved them all. Lewis Temple, an escaped slave who started a thriving shorefront business in New Bedford, invented a toggle harpoon that became the industry standard, impossible to dislodge from a hooked whale. The six-man boats Yankee whalers used to pursue the leviathans were considered the finest in world: sleek, narrow, brightly painted, thirty feet long and six feet wide, and equipped with a sail and oars. On board the whaling mother ships were state-of-the-art "trying pots" to boil whale oil out of the blubber and dozens of barrels to store it in so that the whaler never had to return to port until it was full.

The whaler *Charles W. Morgan*, in 1841, a model of economic efficiency. Whalers could make $50,000 from the baleen extracted from just a single bowhead whale.

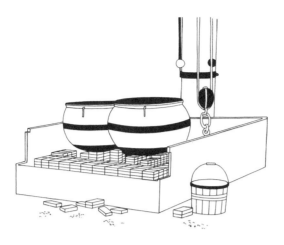

The trying pots on a whaler, in which the blubber of the whales—cut down into thick squares called "Bible leaves"—was melted down for the oil that lit the lamps of America. Whaling was sustainable before the Civil War, concentrating on bull whales, which harbored four times as much oil as females.

A whaling trip was a $50,000 investment, and a risky one at that; sooner or later, 40 percent of the ships did not return. But profits could be tremendous. By the 1830s, it was whale oil that kept the lights burning and the wheels turning in America. Spermaceti, a liquid wax drawn from the massive heads of sperm whales, made the best candles ever: smokeless and odorless, with a high melting point and a low freezing point. It burned brightly in the household lamps of the rich, in city streetlights, in lighthouses, and in the headlights of the new trains thundering out across the West (see page 61). More common whale oil, extracted from the blubber, was an invaluable asset in the emerging industrial economy of the North, greasing the spindles in the cotton and wool mills and the gears of the trains and the steamboats. It made superior soaps, was ideal for softening leather and thickening paint, and worked as a cleaning agent, a cosmetic, and a medicinal ingredient.

Many men signed on for one whaling voyage; few did for a second one—unless compelled to do so by what they owed the shipowners, who deducted every possible expense from their meager earnings. The common seamen's quarters were described by one denizen as "black and slimy with filth, very small and hot as an oven." Its denizens ate salted horse and pork, were whipped with a cat-o'-nine-tails for infractions, and were tormented by vermin drawn to the bloody, greasy ships. No matter how much they scrubbed the decks, their homes at sea stank of cooked whale blubber; whaling ships could be smelled downwind long before they were seen.

"A more heterogeneous group of men has never assembled than in so small a place as is found in the forecastle of a New Bedford sperm whaler," one observer noted, and it was true. Whaling men came from everywhere: from Polynesia and Tahiti, Portugal and the Azores and Cape Verde, Peru and Colombia, and the West Indies. They included Maori from New Zealand; Native Americans from Cape Cod and Martha's Vineyard, renowned for their ability as harpooners; and former slaves, escaped from their servitude on southern plantations.

When the famous cry of "Thar she blows!" sounded from the crow's nest, a hundred feet above the deck, the men scrambled to launch the whaleboats and sailed toward their prey, shouting, "A dead whale or a stove boat!" What followed was a terrifying life-and-death struggle that could easily end with a whaleboat smashed or bitten to pieces by a wounded whale—and the whalers with it. Once the harpoon was in, the men might be pulled by the infuriated animal at twenty miles an hour across the open sea in what was called a "Nantucket sleigh ride." The sailors dumped water on the harpoon line to keep it from bursting into flames as it swiftly unwound, and stood ready to chop it loose with a hatchet if it snagged and started to pull them all under.

Once the whale was finally killed, it had to be towed back to the ship while the men beat off swarms of sharks trying to devour their catch. When it reached the mother ship, all hands descended upon the carcass, chaining it up on the starboard side and peeling it like a gigantic apple with razor-sharp spades, knives, and axes—knowing that one slip could plunge them into the frenzy of sharks gathered below.

They extracted every ounce of value they could find in the gigantic mammals. From the head of the baleen whale they pulled its feeding plates, the plastic of the nineteenth century, used to make buggy whips and parasols and to torture the ribs of countless women as corset stays (see page 107). From deep in the intestines of the sperm whale they scooped out a fragrant gray substance known as ambergris that made the perfume the ladies wore.

Whaling would hang on as an industry into the twentieth century, mainly due to the demand for those baleen corsets. But it was largely finished after the 1850s in the United States, thanks to Confederate raiders that devastated the North's whaling fleet in the Civil War and, especially, a new source of oil, bubbling up from the ground in western Pennsylvania (see page 163). In 1927, the last Yankee whaler left New Bedford, and the terrible romance of the hunt departed with it.

THE GENIUS DETAILS

American whaling's most profitable year was 1853, when eight thousand whales were killed to produce 103,000 barrels of sperm oil, 260,000 barrels of other whale oil, and 5.7 million pounds of baleen, which generated total sales of $11 million.

Whaling crews consisted of sixteen to thirty-seven men. They included a captain, three or four ship's mates, boat steerers, harpooners, a blacksmith, a carpenter, a cook, a cooper, a steward, and foremast hands.

A whaling captain would commonly get a "lay," or cut, of $\frac{1}{8}$ of all profits, while an ordinary crewman would get as little as $\frac{1}{350}$.

Herman Melville had shipped out on the whaler *Acushnet* in 1841 at the age of twenty-one. His book *Moby-Dick* was inspired in good part by the only known case of a whale attacking a mother ship, in 1820.

BLACK GOLD
THE OIL RIG

They knew the oil was there, all over America. In many places it would simply seep up out of the ground, and it was used by Native Americans for decoration and as an insect repellent, a salve, a purge, and a tonic. They would soak it up in blankets or skim it right off the ground. White settlers found it appeared when they tried to dig out salt licks. They retrieved it to lubricate their sawmill machinery or to sell as bottles of patent medicine, but really, what other use was there for it?

One of those snake oil salesmen, an enterprising Pennsylvania merchant named Samuel Keir, discovered a process to extract kerosene from crude and in 1851 began selling this "carbon oil" to local miners for their headlamps. With whale oil, the preferred source of illumination in America (see page 159), growing more expensive, Keir set up the first oil refinery in Pittsburgh.

Now there was a demand, and one Edwin Drake, a former train conductor, provided the supply in 1859, building the world's first modern oil well in the small town of Titusville in the northwestern corner of Pennsylvania. The Chinese had been drilling with bamboo since at least the fourth century, and Europeans had been digging oil wells by hand in Poland and Romania a few years before Drake, but he was the first to use a steam engine to power his drill and the first to drill down through lengths of iron pipe, so as to keep the borehole from closing up.

After months of futility, Drake hit pay dirt near the end of August in 1859. Soon he was pulling up four hundred barrels a day, pumping the first load of oil into a convenient bathtub before he was able to build enough barrels to hold it. Drake set off the world's first oil boom, with people pouring into the country around Titusville to start drilling.

Within another generation, oil had largely put paid to the American whaling industry and had set a young refiner named John D. Rockefeller on the road to amassing perhaps the greatest fortune in US history. But it had not yet assumed the central place it would in the world economy, in part because no one was sure just how much of the stuff there was.

By 1902, 285 oil wells were clustered on the little Texas hill of Spindletop during the state's runaway oil boom.

Spindletop's salt cone dates back to the Jurassic era. Its name may have come from the way in which a grove of trees on top of the hill seemed to "spin" when heat rose from the prairie around it. Before the oil was discovered, many ghost stories were set around Spindletop, and St. Elmo's fire used to play about the mound.

The population of Beaumont rose from eight thousand to over sixty thousand in the space of a year.

Texas's state antitrust laws banned John D. Rockefeller's massive, would-be monopoly, Standard Oil, from Spindletop, allowing the rise of several new major oil companies, including Texaco and Gulf Oil.

The city of Houston quickly became the country's oil capital, and the United States surpassed Russia as the world's leading oil producer.

This would change dramatically early in the new century, when a veteran wildcatting team headed by brothers Curt and Al Hamill drilled into a salt dome near Beaumont, Texas, called Spindletop.

By this time, offshore oil wells had been drilled in tidal zones all along the nearby Gulf Coast of Texas and Louisiana. Pattillo Higgins, a half-mad, irascible local "oil prophet" who had educated himself in geology, had identified the dome as a likely oil deposit at a time when most professional geologists thought the area was too sandy to contain oil. He in turn had called in an immigrant Croatian geologist named Anthony Lucas (born Antun Lučić), who confirmed his suspicions. But lacking the money to exploit Spindletop, Higgins was forced to sell out to a couple of big-oil money men from Pittsburgh named John H. Galey and James M. Guffey. It was Galey and Guffey who hired the Hamills, who they knew had already perfected a superior rotary drill. Now the Hamills hit upon a revolutionary new technique. To keep the sandy Gulf Coast earth from clogging up their works, they bought a small herd of cattle, let it cavort in a waterhole, then used the mud on their drill to clean out the sand. It worked, and mud would become a universal addition to the drilling process, not only washing debris out of the borehole but cleaning and cooling the drill bit and helping prevent gushers as the drill descended.

Even with the mud, the Hamills's work was grueling and dangerous. The layers of rock between the sand wore away their drill bits. Sudden pockets of natural gas threw boulders up out of the earth and damaged their machinery until they kept the drill working around the clock, afraid another blowout would end their whole enterprise.

On January 10, 1901, just after the Hamills had put in a new bit, mud began to boil up out of the ground with such force that it pushed the pipe up through the derrick—something they had never seen before. As they watched, incredulous, the pipe knocked the crown block off the derrick, then fell to the ground "like crumpled macaroni," taking out the boiler smokestack. Then came a deafening roar and a geyser of mud that coated the men, followed by

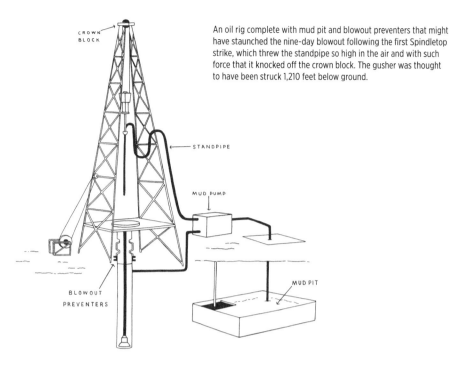

CROWN BLOCK

STANDPIPE

MUD PUMP

BLOWOUT PREVENTERS

MUD PIT

An oil rig complete with mud pit and blowout preventers that might have staunched the nine-day blowout following the first Spindletop strike, which threw the standpipe so high in the air and with such force that it knocked off the crown block. The gusher was thought to have been struck 1,210 feet below ground.

more mud and blue gas flames that shot out of their six-inch drill hole with "a roar like a cannon." When it was over, the team crept closer, peering down into the hole.

First they could hear it, gurgling beneath the earth. Then they could see it—oil forced up with such pressure that it frothed as it came. As the men ran for their lives, the oil blew a new volley of rocks sky high, then blasted out into a 160-foot geyser, twice the height of the derrick. It rained down on the Hamills, and on the crowds who flocked to see it, every time the wind changed. The gusher produced nine hundred thousand to one million barrels before it was finally capped—one of the most productive single oil wells ever found. By itself, it increased US oil production by 50 percent and world production by 20 percent. Its daily production was twice that of all the oil wells in Pennsylvania, then the nation's leading oil state, and six times that in all of California.

And save for forty barrels (of forty-two gallons each), it was all wasted, spilled out on the surrounding ground, where a few days later, most of it burst into a raging fire. But by the end of the decade, more wells on Spindletop would triple total US oil production. That single Texas salt dome would still be producing as much as twenty-one million barrels a year by 1927 and would give up a total of 153 million barrels by 1985. Its abundance would lead wildcatters to confirm that oil was everywhere: all over Texas, and Louisiana, and Southern California, and even under the ocean. It was the perfect energy source for the new internal combustion engines that were soon powering automobiles everywhere.

For better or for worse—and it would become a highly problematic addiction—oil was here to stay.

THE ELECTRIC LIGHT

A s America's economy raced toward becoming the largest in the world by the 1870s, there was only one thing it lacked: light. Light for its factories now churning twenty-four hours a day, for the streets and homes of its burgeoning cities, and for the libraries where its young men and women read on, long after their punishing work-day was done.

Where was the light to come from? Only so many whales could be caught and boiled (see page 159). Kerosene distilled from crude oil (see page 163) gave off a flickering, smoky light, while gaslight was dim, sinister, and dangerous.

The answer lay with an illumination that had first put America on the scientific map, and with the nation's first wizard. The potential of electricity had fascinated the Western world since Ben Franklin's experiments in the 1750s. Its power had been appropriated to invent the telegraph (see page 67), the telephone (see page 74), and the phonograph (see page 218)—three world-changing devices Thomas Edison had already played a key role in either inventing or improving by 1878, when he decided to use electricity to light up the world.

The sticking points were how to transmit electricity to a dozen different lights—or a million—at the same time and how to get it to glow in a lightbulb without immediately burning out the filament. But beyond his brilliant and unorthodox mind, his tireless energy for experimentation, and his ability always to look to the practical applicability of any new invention, Edison had another ace in the hole: Menlo Park, New Jersey, a homey complex of laboratories, machine shops, library, and living quarters—the first in a long line of American industrial research centers. It took up two city blocks and included a miniature locomotive Edison liked to drive at up to 40 miles an hour around a 2.5-mile track.

There "the Wizard of Menlo Park" was able to attract some of the most talented young scientists, engineers, and craftsmen from around the world. Together they would invent the entire system of electrical energy almost from scratch, starting with a new dynamo with more than 90 percent efficiency—over twice as much as the best dynamos then extant. Next came a new, handcrafted glass lightbulb that sealed in an almost perfect vacuum.

A traditional lightbulb, with a tungsten filament.

The right filament proved a sticking point, requiring months of painstaking, dangerous work, with Edison temporarily blinded in one accident. The scientists tested everything from coconut hair to sassafras to a spider's thread and a strand plucked from an assistant's beard before finally settling on carbonized cardboard that lasted for forty-five hours without burning out. A few months later, they would switch to a carbonized bamboo filament, which lasted for 1,200 hours.

By New Year's Eve 1879, Edison was ready for a typically canny public demonstration. All that day and evening, the special trains kept bringing people out to Menlo Park from New York, despite a growing snowstorm. Edison and his men lit up the little town from his lab to the rail station, four hundred lightbulbs in all, giving off what a reporter called "a bright, beautiful light like the mellow sunset of an Italian autumn." As the old year sank away and the light grew in the darkness, delighted gasps of "Marvelous!" and "Wonderful, wonderful!" rang from the crowd.

"We will make electricity so cheap that only the rich will burn candles," Edison informed them.

Still, installing the first electrical system, in Lower Manhattan, would mean designing and manufacturing six twenty-seven-ton dynamos; fourteen miles of insulated, underground wiring; a steam engine with four coal-fired boilers; electrical motors, meters, switches, sockets, fuses, and more. Edison had to find and buy a location for his central electrical station, market and advertise his new invention, sell it to the public and municipal authorities, and put up a majority of the capital—something that would cost him most of his shares in the new Edison Electric Illuminating Company. Much of the work he did himself, from jumping down into trenches in the streets of Manhattan to check the wiring, to hosting the mayor and the board of aldermen at a Menlo Park reception catered by Delmonico's.

Finally, on September 4, 1882, Pearl Street Station, the first central power plant in the world, went on line in Manhattan, sending 110 volts of direct current to bulbs in 450 lamps belonging to eighty-five customers—including several of the city's leading papers, another perfect Edison touch. The *New York Times* reported how the new light provided a glow that was "soft, mellow and grateful to the eye, and it seemed almost like writing by daylight."

Within a year, Pearl Street Station had 513 customers and was lighting ten thousand lamps. The first electric sign appeared on Broadway in 1892, and by 1913 there were over a million lights along New York's Great White Way. By 1904, the *Times* would be there, too, in a tower 375 feet high, and "said to be visible from eight miles away," according to historian Jim Traub, "an 'X' that marked the center from which the great, glowing city radiated."

Here was the perfect synergy of the twentieth century, a transportation hub anchored by a skyscraper, housing a newspaper, and all of it lit to the skies with electricity. Nobody knew it yet, but Times Square marked the start of the decades-long shift of the city's main reason for being from manufacturing to the new business of entertainment.

Electricity was the third, vital element necessary to create the modern, urban experience, along with steel-frame construction (see page 135) and Elisha Graves Otis's safety elevators (see page 127). Electricity lit homes and offices. It powered the underground and underwater commuter trains (see page 21) enabling hundreds of thousands, then millions, of individuals to be assembled in, and dispersed from, the city in record time—thereby inventing the suburbs, too.

Electric light could seem terrifying, even surreal—"Squares after squares of flame set and cut into the ether," as Ezra Pound would write. But "incandescent lighting," as the cultural historian David Nasaw noted, "transformed the city from a dark and treacherous netherworld into a glittering multicolored wonderland."

Where murky gaslight had only exaggerated the terrors of the city, clean, dazzling electricity made people—a whole new class of people, single, independent office workers, male and female, with some discretionary income in their pocket—want to linger after work. To meet their needs, entertainment became an industry as never before, giving birth to fantastic, electric-powered amusement parks (see page 229) and sumptuous new palaces where electricity ran the projectors that spewed forth the new century's boldest new art (see page 239). The Wizard, unsurprisingly, would soon turn his energies to inventing the movies.

THE GENIUS DETAILS

Other substances tested for the lightbulb filament included rags, grasses, flour paste, leather, macaroni, fishing line, pith, cinnamon bark, eucalyptus, turnips, gingerroot, cedar shavings, hickory, maple, cork, twine, celluloid, flax, paper, vulcanized fiber, and cork.

Soon after lighting up Menlo Park for New Year's Eve 1879, Edison and his men lit up the mansion of one of his backers, J. P. Morgan. It required 385 bulbs, as compared to just 400 for all of Menlo Park.

Beginning in 1928 in New York City, Consolidated Edison gradually replaced Edison's direct current with alternating current and completed the switch in 2007. Older buildings—and the New York subways—were fitted with converters to change alternating current, the only kind now generated by ConEd, into direct current for their internal use.

Moses Farmer, an electrical pioneer in Salem, Massachusetts, lit his home with incandescent bulbs in 1859, making it the first electrically lit home in history. But the bulbs quickly burned out.

TAMING THE COLORADO
THE HOOVER DAM

T he river had always seemed to have a mind of its own. It ran as wild as any water on earth, and each year its rampant flooding killed hundreds of people and drowned thousands of acres of farmland. Every attempt to master it had failed. It had smashed apart a canal designed to channel its waters into California's Imperial Valley, causing massive damage and creating a 150-mile-wide, 60-foot-deep lake known as the Salton Sea. Geologists predicted that if something wasn't done to divert it back to its original course, it would plow itself a canyon a mile deep and hundreds of miles long.

The Colorado would have to be dammed.

It was as complicated and forbidding a project as any that had confronted Americans in the West, including cutting the Transcontinental Railroad through the Sierras and bridging the Golden Gate (see pages 48 and 141). President Herbert Hoover ordered construction sped up to begin in May 1931 in order to provide jobs in the face of the growing Depression. But a planned company town had not been built yet, and many workers and their families were forced to live in a wretched work camp known as Ragtown. By August they had gone on strike over the bad water, the awful food, a sudden wage cut, and work in weather that killed sixteen men from heat prostration.

A consortium of six construction companies was hired to do the work, and Frank "Hurry Up" Crowe, a tall, stately, fifty-year-old Quebecois who had been building dams all over the West for thirty years, was brought on as superintendent. Crowe's motto was "To hell with excuses—get results," and he lived up to it. He broke the strike by summarily firing everyone, then rehired the men he wanted, housing them in the model town of Boulder City that the Nevada authorities had finally gotten around to building. Even there he ran roughshod over his workers, but Hurry Up Crowe at least lived up to his own punishing work ethic. Every detail required—and got—his close attention.

What he saw was the incredibly variegated set of tasks necessary for building his dam. Just to begin it, Crowe's men had to divert the mad river from Black Canyon, on the border of Nevada and Arizona, where it tore through at up to 175 miles per second. This meant building an enormous cofferdam and diverting the river into four tunnels that had to be drilled and dynamited into the canyon walls. To build the cofferdam, "high-scalers," suspended from the top of the canyon by rope, would remove

The magnificent Hoover Dam, which today lights the Southwest.

In his haste to complete the dam ahead of time, "Hurry Up" Crowe nearly caused an even greater disaster. Pushed by their superintendent to finish the dam's foundation, the men simply left some fifty-eight cavities in the bedrock when they realized it would take "too long" to fill them with grout. Repeated water seepage revealed this defect, which was rectified only by nine more years of work filling in the grout curtain from 1938 to 1947.

One hundred and twelve people died building Hoover Dam. The consortium of construction companies listed forty-two of them as victims of "pneumonia," but this was a cover-up of deaths caused by carbon monoxide, due to carelessness in the rush to build the diversionary tunnels.

Eventually, enough water to fill fifteen swimming pools every second would pass through the dam's seventeen generators, producing 2,080 megawatts of power, more than enough to repay its construction costs ($49 million, about $833 million in today's dollars) and its yearly maintenance expenses.

loose rock with drills and dynamite to prevent water seepage, in between swinging about spectacularly to show off for the tourists. Inside the tunnels temperatures reached 140 degrees, and one man after another keeled over, some of them dying of carbon monoxide poisoning from the drilling machinery. But the tunnels were finished eleven months ahead of schedule.

Next, the men had to dig as deep as 150 feet into the bedrock, clearing out 1.5 million cubic yards of material—twice the amount displaced by digging the Panama Canal—and filling in any natural cavities they encountered with a "grout curtain." Then came the enormous logistics challenge of how to run the worksite.

"We had 5,000 men jammed in a 4,000-foot canyon," Crowe later explained. "The problem . . . was to set up the right sequence of jobs so they wouldn't kill each other off."

Crowe employed any number of techniques and devices he had improvised at previous dam sites. He designed and built the most sophisticated cable system ever seen, used to deliver both concrete and workers alike around the job site. It swung huge, steel buckets—patented by Crowe, transported on special railcars, and capable of carrying twenty short tons of concrete apiece—to pour into the dam.

Still, there remained the problem of the dam's sheer size, 45 feet wide at its rim and 660 feet thick at its base. It would require 6.6 million tons of concrete, enough to lay a 4-foot sidewalk around the earth at its equator. Done in a single pour, the amount of concrete needed would have taken 125 years to cool and harden—and would have cracked in the process, rendering it useless.

The dam's designer, John L. Savage, had the solution to this: 230 wooden box molds, each equipped with 1-inch pipes and thermometers—582 miles of pipe in all. The concrete was poured into each mold, 5 feet deep; then refrigerated water was pumped through the pipes to cool it down. When the thermometers indicated that it was cool enough, grout was pumped into the pipe sections—thereby further strengthening the dam—and the next section was poured. Sample cores taken from the dam in 1995 showed that "Hoover Dam's concrete has continued to slowly *gain* strength" (my italics), and it will continue to do so for years to come.

The great dam was finished in 1936, two years, one month, and twenty-eight days ahead of schedule, and on its completion it was the largest concrete structure in the world, as well as the largest dam and greatest single generator of electricity. The power it makes and the water of its reservoir make possible the continuing existence of civilization as we know it in Southern California and throughout the Southwest.

Behind its concrete walls, the dam created a huge new body of water, Lake Mead, 500 feet deep and 110 miles long. It continues to serve as the largest reservoir in the country and is a highly popular recreation site.

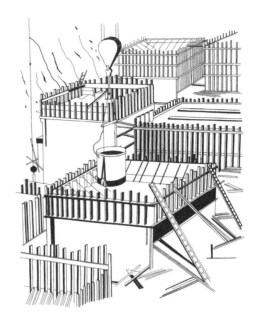

Some of the 230 wooden box molds the 6.6 millions tons of concrete in the Hoover Dam were poured into. Cooled with water pumped in through 582 miles of pipe, then filled with more concrete, the dam is *still* growing stronger, and will do so for years to come.

Beyond all its practical uses, however, the dam is also a work of surpassing beauty. Architect Gordon B. Kaufmann's monumental art deco architecture for the roadway, the water intakes, and the generating plant at the top of the dam are perfectly complemented by the motifs of the Navajo and Pueblo tribes of the region—rain, water, lightning, clouds, lizards, serpent, birds, and the surrounding mesas—that designer Allen Tupper True embedded in the plant's walls and the terrazzo floors. A. T. True—much like the designers of Pennsylvania Station or the Coney Island parks—also made the practical mechanics of the dam, the giant turbines and pipes, works of art. This was the work of a people supremely confident, even at the depths of the Depression, in its own ability, and its future.

THE TENNESSEE VALLEY AUTHORITY

N

o region of the country was harder hit by the Great Depression than the Deep South, which had never fully recovered from the devastation of the Civil War. But with the Depression would come opportunity at a bend in the river.

A tortuous twist in the Tennessee River, at a place called Muscle Shoals, Alabama, sent water plunging 140 feet over the course of thirty miles—and wherever water rushes and falls with such ferocity lies power (see page 171). During World War I, the US government built two plants there to process nitrates for use in munitions. They were supposed to be powered by a hydroelectric dam (named for President Wilson), but the war ended before the dam could be completed.

It was another setback to a region where the average annual income would still be just $639 by the end of the 1920s, and many families made as little as $100 a year. One-third of the population suffered from malaria. The soil was played out after decades of overuse, and deadly floods from the unruly Tennessee constantly swept away family farms. In their desperation for fuel or new land to farm, the locals chopped or simply burned down 10 percent of the area's once capacious forests every year.

Private utilities saw no reason to provide this indigent area with electricity. While local cities were over 90 percent electrified by 1929, only one in ten rural families had electric power, the rest still living much as they had a century before.

Then manna seemed to drop from heaven. The unfinished dam project attracted the attention of Henry Ford, still a name to conjure with in American life. Ford brought his friend Thomas Edison down to Muscle Shoals near the end of 1921 and offered to transform the region.

"I will employ one million workers at Muscle Shoals, and I will build a city seventy-five miles long at Muscle Shoals," he promised.

The Wilson Dam, one of forty-seven dams built in the vast Tennessee Valley Authority system.

The turbines of the Wilson Dam were works of art. Their power helped transform both life and commerce in the South.

Who could doubt the man who had brought America the Model T (see page 147), especially with the wizard of invention (see page 218) at his side? Giddy Alabamans immediately started a land boom. But not everybody was convinced.

First among the doubters was one George Norris, an independent-minded Republican senator from Nebraska. Norris had been interested in the potential of Muscle Shoals since the war, even though he'd never been there. Now he went to see the region for himself and thought Ford's offer was big on dreams and short on cash: only $5 million offered for a project that had already cost the government $130 million.

It may seem incredible today, a senator from one party concerned about a region dominated by another party, hundreds of miles from his home state. But then Norris, whom many historians name as the greatest senator in our history, never much cared for the conventions of party politics. One of eleven children, he was raised in poverty after his father died when he was four, and worked his way through law school. Elected to Congress from Nebraska as a Progressive Republican in 1902, he led a rebellion in the House that overturned the autocratic rule of his own party leaders. Elected to the US Senate in 1912, Norris, who looked and acted like a hero from a Frank Capra movie, was one of only six senators to oppose America's entry into World War I. Throughout his long career he sided with whichever party he believed would help him in his ceaseless quest to "repudiate wrong and evil in government affairs."

Norris convinced the Senate to turn down Henry Ford's offer to buy the Wilson Dam. When he returned to Muscle Shoals, he had to bring an armed bodyguard. But he was not about to abandon the area. Twice in the 1920s, Norris got bills passed to finish the abandoned Wilson Dam and develop public power there. Not one but two presidents from his own party vetoed these bills as socialistic.

In 1933, the man and his moment were finally met. Franklin Delano Roosevelt saw a chance to put into place all of the ideas he'd long had regarding public power, planning, economic development, and conservation. FDR pushed the Tennessee Valley Authority (TVA) through Congress during the famous "First Hundred Days" of his administration.

The federal government completed the Wilson Dam—and a dozen others, over the next few years—in what constituted the largest hydropower construction program ever undertaken in the United States. But the TVA was much more than just a utility. Here was the genius of America not in a single, doughty inventor but in a comprehensive development plan for a region that had long lagged behind the rest of the nation. The new dams controlled the constant flooding, made whole rivers navigable, and created beautiful lakes for recreation and fishing. The power was used to bring the cheapest electrical rates in the country to everyone, on the farm or in town. (It also set a standard for rates nationwide, which drove the private utilities mad.) Electricity transformed daily life, bringing industry to the area and electric light, electric washing machines, and electric refrigerators to people's homes. The Department of Agriculture used the nitrates produced at the old war plants to refurbish the soil and set up services for farmers, increasing local agricultural yields. The Forest Service and the Civilian Conservation Corps replanted the trees and made the valleys bloom again.

Trying to supply its workers with reading material, the TVA even started local public libraries throughout the region in stores, post offices, and gas filling stations when necessary. Run by Tennessee librarian Mary Utopia "Topie" Rothrock, the libraries remained even when the dams were finished.

Advocates for private utility companies and limited government fought the TVA all the way to the Supreme Court. But the Court ruled that it was constitutional—and the TVA rewarded George Norris's faith by quickly paying off its cost and becoming self-sustaining. By 1941, the authority was the largest producer of electrical power in the United States, and it would prove invaluable to the war effort, powering the production of aluminum and other vital metals, countless nitrates, and a top-secret program at Oak Ridge, Tennessee, that was part of something called the Manhattan Project.

This was a war effort that George Norris fully supported. Previously an isolationist, he had come out for US intervention after seeing pictures of Japanese army atrocities in China. But then, Senator Norris always did think for himself.

Centered on the Tennessee Valley, the TVA serves customers over eighty thousand square miles, including most of Tennessee, much of Mississippi, Alabama, and Kentucky, and parts of Georgia and North Carolina.

Sixteen hydroelectric dams and a steam plant were constructed by the TVA between 1933 and 1944, and a total of forty-seven dams were built in six different states.

During the Great Depression, the TVA was a great tourist attraction. One thousand people a day came to see the Wilson, Wheeler, and Norris dams.

The TVA set up the Electric Home and Farm Authority (EHFA), which offered low-cost financing to farmers, to allow them to buy electric stoves, refrigerators, and water heaters at affordable prices.

The TVA is still America's largest public power company, with seventeen thousand transmission lines delivering nearly thirty-two thousand megawatts of power to 8.5 million individuals. Its revenue in 2013 was almost $11 billion and its operating income nearly $1.5 billion.

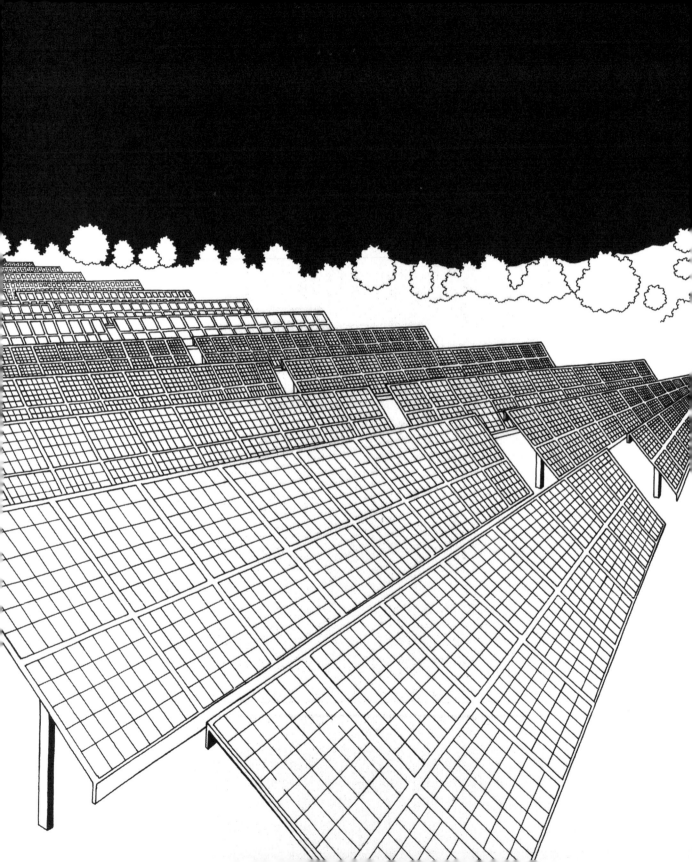

THE SOLAR CELL

T he rates at which scientific discoveries advance into practical application can seem capricious, determined as they are by so many outside factors: the availability of related enabling technologies; government support; or market economics. But one factor that the American experience with invention shows to have always been a boon is the gathering of brilliant minds in one place.

Such a place was Bell Labs, the now legendary New Jersey research center for AT&T that pioneered groundbreaking work in so many fields (see page 81). When its final history is written—many years from now—perhaps the invention of the modern solar or "photovoltaic" cell will be considered its most important breakthrough.

The concept of the solar cell had been around for a long time. In 1839, nineteen-year-old Alexandre-Edmond Becquerel invented the very first such cell, showing how light could be converted into electricity in his father's Parisian laboratory. The American inventor Charles Fritts developed the first solid-state photovoltaic cell in 1883 by coating the semiconductor selenium with a thin layer of gold, but it proved only about 1 percent "efficient" in converting light to electricity.

The next big step forward wasn't taken until Bell Labs engineer Russell Ohl discovered the "p-n [positive-negative] junction" or "barrier" in 1939. This boundary between two different types of semiconductor material would serve as a diode, a circuit that would allow one to send a current of electricity in one direction but prevent it from going in another. Think of it as a sort of subatomic circuit breaker.

While transistors were beginning to transform our world after World War II, Daryl Chapin was working on a seemingly more mundane project at Bell Labs: trying to come up with an energy source to power dry-cell batteries for Bell Telephone in the tropics, where they degraded too fast in the intense humidity. But the only solar cells then available were selenium ones that converted just 0.5 percent of sunlight to energy—less than a tenth of the nearly 6 percent level they had to reach to be commercially viable.

Chapin turned to an old friend, Gerald Pearson, a fellow research physicist at Bell and former classmate of his from Willamette University, who was doing semiconductor work with Bell chemist Calvin Souther Fuller. Pearson and Fuller had already

The energy of the future: a field of solar panels today.

gained insight into how to transform silicon from a poor conductor of electricity to a superior one by introducing impurities into it. Now they discarded Chapin's selenium cell, introduced gallium into silicon, then gave it a nice hot lithium bath.

Voilà! Pearson and Fuller had created the best solar cell yet devised, with 2.3 percent efficiency. But just then a completely different energy source—one that seemed about to make solar energy or, for that matter, any other kind of energy all but irrelevant—barged noisily onto the scene.

In January 1954, "General" David Sarnoff, the bombastic head of RCA, announced that his company had made a dramatic breakthrough: the "atomic battery." At a press conference at Radio City, Sarnoff tapped out "Atoms for Peace" on an antique telegraph powered by his new device.

"Atomic batteries," Sarnoff told reporters, "will be commonplace long before 1980," predicting that they would power "ships, aircraft, locomotives and even automobiles. . . . Small atomic generators, installed in homes and industrial plants, will provide power for years and ultimately for a lifetime without recharging."

Americans were eager to hear that there could be a great new, constructive role for atomic energy in peacetime. The *New York Times* called Sarnoff's predictions "prophetic" and predicted the atomic battery would also fuel "hearing aids and wrist watches that run continuously for the whole of a man's useful life."

"Who cares about solar energy?" crowed the head of RCA Laboratories. "Look, what we really have is this radioactive waste converter. That's the big thing that's going to catch the attention of the public, the press, the scientific community."

"Radioactive waste" was right. The atomic battery ran on strontium-90, one of the most hazardous elements of nuclear waste.

All that pesky radiation stuff would ultimately preclude development of the atomic hearing aid—or, for that matter, the atomic anything. At Bell Labs, meanwhile, Calvin Fuller suspected that a problem with his team's solar energy cell was that the p-n junctions tended to be too far from the surface of the cell for enough sunlight to penetrate. Fuller cut the silicon for the cells into long, narrow strips, then added a smidgen of arsenic and coated it all with an extremely thin layer of boron in a furnace. Eureka! The arsenic created p-n junctions, and the boron kept them very close to the surface of the cell, thereby making it possible to pick up sunlight and distribute 5.7 percent of it as electricity.

On April 25, 1954, just three months after "General" Sarnoff's announcement, Bell Labs announced that its "solar cells," linked together, delivered "power from the Sun at the rate of 50 watts per square yard, while the atomic cell recently announced by the RCA Corporation merely delivers a millionth of a watt" over the same area. The *New York Times* announced "the beginning of a new era, leading eventually to the realization of one of mankind's most cherished dreams—the harnessing of the almost limitless energy of the sun for the uses of civilization."

Not just yet. The low cost and ready availability of other fuels constrained the further development of solar cells for decades. But constant improvements on the photovoltaic cell, plus the need to combat global climate change, have made it a competitive energy source and the most likely energy of the future. Best of all, we won't glow in the dark ourselves.

Far from the original Bell Labs quest to attain 6 percent efficiency for solar cells, the most efficient photovoltaic cells today reach 46 percent efficiency, or more than twice the 20 percent efficiency of internal combustion engines. Most commercial solar panels in use today, though, routinely achieve 15 to 18 percent efficiency, with some above 23 percent.

Bell Labs' announcement of its new solar cell in 1954 included a press conference in which the cell fueled a twenty-one-inch Ferris wheel and powered a radio transmitting a song.

The new solar cells were of immediate interest to the US military, which used solar panels to help fly the Vanguard 1 satellite. Solar panels would go on to play a key role in powering geostationary communications satellites.

In 1954, the world had less than a single watt of solar cells capable of running electrical equipment. By 2004, over a billion watts of solar power were in active use, and solar energy was a $3 billion to $4 billion industry in the United States alone.

As of early 2015, some 784,000 American homes and businesses operated on solar power, and there were over 174,000 people employed in the solar power industry.

THE PENNSYLVANIA RIFLE

Boys learned to hunt from a young age on the Pennsylvania frontier. Their fathers taught them to steady their guns on fallen logs or along tree branches, firing from a sitting or lying position as well as standing up. The gun they used, often as tall as they were, was the most deadly weapon in the world at a distance: the Pennsylvania rifle.

Also known as the Kentucky rifle, or the long rifle, it had a particularly American pedigree. Like so many of its users, the Pennsylvania rifle was a hybrid, a perfect combination of German engineering and English style. It would play a vital role in giving America its freedom.

German gunsmiths had been cutting spiraled grooves in their muskets since at least 1450 to prevent gunpowder residue from building up and fouling the pieces. They discovered an added benefit: the spirals propelled the gun's lead ball with the range and the accuracy of an arrow, something no musket—with its smooth barrel—could do. Over the next two hundred years, they added an advanced flintlock firing system and a butt that enabled shooters to hold these "rifles" against their shoulders and aim them more accurately than ever.

By 1719, German gunsmiths were well established in Pennsylvania—thanks, ironically, to the religious tolerance practiced by the colony's pacifist Quakers. Soon they could be found all along the Great Wagon Road, where wagon teams (see page 5) of pioneers pushed west and south into Kentucky and the Blue Ridge mountains, and from there out across a continent. They usually made their rifles from scratch, a trade that required them to master a half dozen crafts, as blacksmiths, whitesmiths, brass and silver workers, carvers, engravers, and finishers of wood. Their guns would become works of art, and over the years they also developed a crucial innovation, adding the longer barrels of English smoothbore hunting rifles, called fusils.

Unlike the peasants dragooned into European armies who had never used a gun before in their lives, Americans on the frontier had been shooting from the time they were at least thirteen. Knowing how to fire a rifle could be a matter of survival, and everything had to be as light as possible to carry on their hunting and exploring forays into the wilderness. The barrels of Pennsylvania rifles, now routinely four feet long, burned black powder more efficiently, which allowed their shooters to fire lower-caliber ammunition that was still not only more accurate but more deadly than heavier musket balls, thanks to the added velocity.

Compared to modern rifles, Pennsylvania rifles were not very accurate. But in the eighteenth century they were much more deadly than the musket. In a 1920 experiment, a Pennsylvania rifle hit a target five times out of ten at a distance of three hundred yards—or three football fields—while a musket hit it only once. Where a musket fired by a European soldier of the line could kill at a hundred yards (and was not truly accurate at any distance), a trained marksman with a Pennsylvania rifle could kill easily at two hundred yards—maybe even twice that distance.

George Washington saw the advantage such marksmen could provide from the outset of the Revolution. By 1777, he had placed five hundred of the Continental Army's best riflemen under the command of Colonel Daniel Morgan, a frontier character who liked to call Washington "Old Horse" and himself "the Old Wagoner." The son of Welsh immigrants to New Jersey, Morgan had left home at sixteen after a fight with his father and settled in Winchester, Virginia. He loved to drink, gamble, and fight, could barely read, and possessed a volcanic temper. But he was an easy disciplinarian and a beloved father figure to his men. Like them, he refused to stand on military ceremony and preferred to dress in buckskin shirts and leggings, moccasins, and coonskin caps, or full Indian gear.

Despite his homespun ways, Morgan was a military prodigy—according to one historian, "the only general in the American Revolution, on either side, to produce a significant original tactical thought." He also had a grudge to settle. Working for the British in the French and Indian War, the Old Wagoner was shot through the cheek and neck, leaving

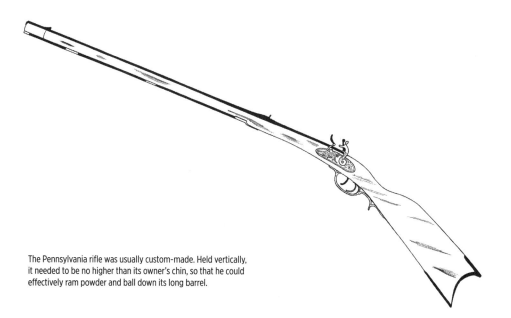

The Pennsylvania rifle was usually custom-made. Held vertically, it needed to be no higher than its owner's chin, so that he could effectively ram powder and ball down its long barrel.

his face permanently scarred. Nonetheless, he had received five hundred lashes for striking a British soldier. He liked to joke that his flogger had miscounted and that "the king owes me a lash"—an account he would settle with interest.

During the Saratoga campaign, the turning point of the Revolution, Morgan's riflemen systematically shot down as many of British general "Gentleman Johnny" Burgoyne's Native American guides as possible, speeding their departure and leaving the British without useful reconnaissance. On September 19, 1777, as Burgoyne's men started to cross a fifteen-acre clearing known as Freeman's Farm, Morgan's riflemen fired from the woods, instantly killing or wounding every British officer in the front line.

This sort of warfare, shooting officers and specialists from cover, was considered "assassination" by the British and little better than a war crime. But Morgan and his men were not about to give it up. When Burgoyne led 1,500 of his best men out into an uncut wheat field at the Battle of Bemis Heights on October 7, Morgan's sharpshooters shot Gentleman Johnny's horse from under him and sent balls whistling through his hat and waistcoat. When Burgoyne's best officer, General Simon Fraser, rallied his troops, Morgan told his men, "That gallant officer is General Fraser. I admire him, but it is necessary that he should die. Do your duty."

Tim Murphy, the twenty-six-year-old son of Irish immigrants to Pennsylvania, climbed into a tree and took aim. His first shot severed the crupper of Fraser's saddle, the next grazed the mane of his horse. His third shot hit the gallant general in the chest, mortally wounding him and halting the British advance. Saratoga would be a crushing victory, one that brought France into the war and eventually assured the rebels their independence, thanks in no small part to a piece of uniquely American technology.

The firing mechanism on a Pennsylvania rifle demonstrates what highly prized tools these guns had become by the 1770s, often as ornately decorated as valuable pieces of furniture, with baroque and rococo inlays and leaf scrollwork.

MOBILE WARFARE
THE REPEATING RIFLE

The American Civil War would last four years, but its weapons and tactics would traverse a century and a quarter of warfare, from the field at Waterloo to the trench combat of World War I to Hitler's blitzkrieg. At the start of the conflict, armies maneuvered very much as they had in the Napoleonic Wars, striving to bring massed infantry fire to bear at as close a range as possible. Troops still used muzzle-loading muskets or rifles, with which they might get off one or two shots a minute until the lead residue fouled the barrel.

Since at least 1808, European gunsmiths had been inventing breech-loading guns with "self-contained" cartridges—that is, guns loaded at their middle or back, with cartridges that included the bullet and its propulsive fuel all in one package, so that the shooter didn't have to ram powder and shot separately down the barrel. In 1848, New York inventor Walter Hunt (see page 97) introduced his "Volition Repeating Rifle," which boasted the first practical, lever-action repeating mechanism, and his "Rocket Ball," caseless ammunition that was loaded with a tubular magazine. But Hunt's prototype was a fragile beast, and by the time the Civil War broke out, both sides were still relying on antiquated muskets and rifles.

Christian Miner Spencer would change all that. The grandson of a Revolutionary War veteran, he was first apprenticed to a silk manufacturer, then worked as a machinist for Samuel B. Colt and some of the other gunworks clustered throughout the Connecticut Valley—shops that bore the names of men like Oliver Winchester, Eliphalet Remington, and Smith & Wesson. By the late 1850s, he was working eleven hours a day, six days a week, perfecting a machine he'd patented . . . to stick labels on spools of silk ribbons. In what spare time he had, though, he labored at a design for a better breech-loading rifle, and he had produced and patented one by 1860.

In 1861, Spencer got an audience with the navy. Lincoln's secretary of the navy, Gideon Welles, was looking for a lighter, rapid-firing gun for his sailors to use. The navy put Spencer's rifle through its paces, burying it in sand and immersing it overnight in salt water. Still, it fired 251 times, with only a single misfire. Impressed, Welles put in an order for seven hundred guns.

Lincoln's chief of ordnance, Gen. James Ripley, however, refused to authorize it. Though Lincoln had ordered him to seriously consider new inventions from citizens, Ripley, like many older officers, considered all breech-loading weapons to be "newfangled gimcracks" and worried that they would inspire soldiers to fire "too rapidly."

Spencer took to the field, demonstrating his weapon for any Union commander who would see him. Gen. Ulysses S. Grant attested that it was "the best breechloading arms available." In January 1862, the innovative Col. John T. Wilder, a nationally known inventor himself before the war, armed his "Lightning Brigade" of mounted infantry with Spencer rifles, and the next year they swept all before them in the campaign that took the key city of Chattanooga. On the strength of such trials, Spencer eventually wrangled an appointment in 1863 with none other than President Lincoln, who was fond of watching new weaponry being tested in the Civil War and when possible would do the job himself at a makeshift firing range of old lumber just south of the White House. (He was "a very good shot, [who] enjoyed the relaxation very much," according to one of his secretaries.)

When Spencer brought his repeating rifle into the White House, Lincoln asked him to disassemble it "and show me the inwardness of the thing." The president was immediately impressed with how quickly Spencer was able to take his gun apart and reassemble it, needing only a screwdriver. He invited the inventor to shoot it with him the following day at another makeshift range out on the Washington Mall. There, firing from forty feet away, the president hit the bull's-eye with his second shot. The next day he was back out firing the Spencer again.

Gen. Ripley was reassigned to inspecting forts. Soon some 144,000 Spencer rifles, including a smaller, lighter version called a carbine, were pouring out of factories.

The Spencer was not the only breech-loading rifle or carbine. The Sharps rifle was considered a better weapon for snipers, with a longer range, but it still fired just

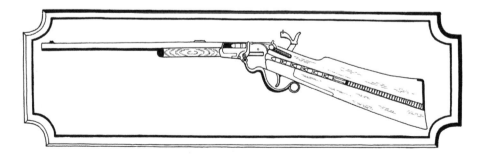

A Spencer seven-shot repeating rifle, complete with Blakeslee cartridge box to be loaded through the stock.

one shot at a time. The Henry rifle had a larger magazine but a shorter range and less punch. The Spencer was the perfect weapon in that it combined all the best features of other carbines and rifles. It was cheap and easy to make, very easy to use, and incredibly durable, almost never breaking down and proving to be generally waterproof. The carbine version (also tested by Lincoln!) was just eight and a quarter pounds and thirty-nine inches long, short enough not to interfere with the cavalryman's arms or tangle in his horse's legs. Its seven-round magazine was spring-loaded from the back of the stock, allowing even green troops to fire off thirty .56-caliber rounds per minute with a range of two hundred to five hundred yards. Reloading was facilitated by the Blakeslee cartridge box, which held six to ten tubes of seven cartridges each that one could simply pop into the Spencer as if he were loading a giant stapler.

With the war in the east bogging down into the sort of murderous trench stalemate that would presage World War I, Gen. William Tecumseh Sherman put into play his theory that cavalry was best deployed across the limitless American landscape by riding to the battle, then dismounting to fight as infantry. The Spencer carbine, light, powerful, and travel-worthy, made such tactics possible. Sherman organized an enormous Western "Cavalry Corps" of some 13,500 men, all of them armed with Spencers. After devastating rebel forces in the battles of Nashville and Franklin, they let rip across Georgia and Alabama, destroying what remained of the Confederate heartland, crushing the command of the previously invincible Nathan Bedford Forrest, and punctuating the end of the war by capturing rebel president Jefferson Davis and vice president Alexander Stephens.

It was all due to the firepower of the carbines they carried, according to their foes, who decried "those damned guns the Yankees can load on Sunday and fire all week."

THE GENIUS DETAILS

In a morbid irony, Lincoln's assassin, John Wilkes Booth, would be mortally wounded in a tobacco farm in Virginia, trying to hold off a contingent of Union cavalry with a Spencer carbine.

Spencer rifles helped Gen. George Armstrong Custer beat back Confederate general Jeb Stuart's assault on the rear of the Union lines at Gettysburg in 1863 and helped Gen. Phil Sheridan's cavalry to smash rebel resistance in the Shenandoah Valley in 1864.

Facing a sudden drop-off in business, the Spencer Repeating Rifle Company went bankrupt in 1868, with its assets eventually bought up by Oliver Winchester, for whom Spencer made further improvements on his rifle.

Christian Spencer would later invent the first practical pump-action shotgun at his new firm, the Spencer Arms Company.

THE GATLING GUN

W hile the invention of the carbine added mobility to the end of the Civil War (see page 185), the Gatling gun contributed something else: overwhelming firepower. Accepted too late by the Union Army to make much of a difference in the war itself, it would play a critical role in other facets of our history, including the preservation of the *New York Times*, the making of one of our most popular presidents—and Winston Churchill.

Richard Jordan Gatling was born into an extraordinary family of inventors, planters, and slaveholders in Hertford County, North Carolina. He patented a rice-seed planter in 1844 when he was just twenty-six and moved to St. Louis to manufacture and sell his new invention. There he turned it into a wheat planter and made a fortune in the rapidly industrializing farm economy of the Midwest (see page 208). Unstoppably inventive and self-sufficient, Gatling would soon invent a hemp-making machine and a steam plow. During the winter of 1845, he contracted smallpox while trapped on an icebound riverboat for two weeks—and after surviving the ordeal went to college and earned a medical degree, mostly for the purpose of being able to treat himself and his family.

The machine gun's deadly predecessor: the Gatling gun.

With his medical training, though, he was also able to inspect the trains full of dead and wounded Civil War soldiers pouring back into Indiana. He could see that only about one-sixth of the dead were killed by actual combat wounds. The rest died from the usual camp diseases of the time: cholera, dysentery, pneumonia.

"It occurred to me if I could invent a machine—a gun—which could by its rapidity of fire, enable one man to do as much battle duty as a hundred, that it would, to a great extent, supersede the necessity of large armies, and consequently, exposure to battle and disease [would] be greatly diminished," he would write, in a rather cold-blooded spirit of philanthropy.

What Gatling did was essentially adapt his old rice planter into a machine gun. Ideas for such "coffee-mill" guns—named for the household device they most closely resembled—had been around for a long time, but Gatling's was uniquely simple, effective, and foolproof. The Gatling gun consisted of six (later ten) rifled barrels mounted around a central shaft. A hand crank both rotated the barrels and dropped cartridges into each barrel from a "hopper" or "stick magazine" mounted above them. The machine did the rest, firing the cartridge and ejecting the casing. The rotation of the barrels allowed each one to cool before the next shot—a vital necessity in Civil War–era weaponry—and if one barrel jammed the rest would keep firing. It required only two men with the most basic instruction to work the gun, which fired as many as two hundred rounds a minute.

Gatling had his gun ready for battle by 1862. Soon he was corresponding with a very interested President Lincoln, who often involved himself with new advances in ordnance. Deployed then, the Gatling gun might have shortened the war by years.

Like many other such inventions, though, Gatling's ran afoul of the War Department's bureaucracy. Specifically, the army was suspicious of his family's slaveholding, southern background. The army would not officially purchase his gun until the Civil War was over. If he *was* a secret Southern sympathizer, he had an odd way of showing it. Individual Union officers bought at least a dozen Gatling guns with their own money, and it proved brutally effective in action, mowing down Confederates trying to break through the Union lines at the siege of Petersburg late in the war.

Yet the greatest contribution of Gatling's gun to the war was off the battlefield. During the bloody Draft Riots in New York City in 1863, an antiwar mob gathered in City Hall Park, determined to destroy the pro-Union establishments of Horace Greeley's *New York Tribune* and the *New York Times*. A major *Times* stockholder was a wealthy Wall Street trader and sportsman named Leonard Jerome, who saw to it that three Gatling guns were posted in the windows of the *Times* and who manned one of them himself. The mob took a good look at them—and decided to attack the *Tribune*. The *Times* went on publishing, and Jerome lived to take his daughter, Jennie, off to England, where she married a British lord and became the mother of Winston Churchill.

The Gatling gun was not technically a machine gun, since it was hand-operated. The first recoil-operated machine gun was the Maxim gun, also invented by an American, Hiram Maxim.

Guns with revolving cylinders and chambers go back at least as far as the British "Puckle gun" of 1718, and the Confederates introduced a five-shot "revolver cannon" at Petersburg. But all such inventions used the same barrel.

"It's the Gatlings, men! Our Gatlings!" Teddy Roosevelt cried, rallying his troops. They and the other American forces surged up San Juan Hill, yelling, "The Gatlings! The Gatlings!" Later, Lt. Parker's Gatlings loosed a deadly fire on Spaniards trying to retake Kettle Hill, killing an estimated 560 of 600 attackers. They took out a Spanish artillery piece and fired some seven thousand rounds into Santiago, hastening its surrender.

The Gatling gun would eventually be made automatic. Its modern, helicopter-mounted descendant, the Vulcan minigun, was widely used by the United States during the Vietnam War, firing 6,000 rounds a minute.

Gatling sold his gun to Colt in 1870. The Gatling gun, alas, proved no more a deterrent to warfare than Alfred Nobel's dynamite and was used mostly by colonial European powers to crush uprisings in their empires.

During the Spanish-American War of 1898, though, Col. Teddy Roosevelt's Rough Riders, two regiments of all-black "Buffalo Soldiers" under Lt. John "Black Jack" Pershing, and other US troops were pinned down by Spanish fire during the Battle of San Juan Hill. The day was saved when a battery of three Gatling guns, brought to the battlefield at the insistence of Lt. John H. "Gatling Gun" Parker, opened up on the Spanish trenches and blockhouses, thus enabling the heroic charges up San Juan and Kettle Hills. The guns later helped to halt a Spanish counterattack and force the surrender of the city of Santiago—as well as saving both Pershing, the commander of all American forces in France during World War I, and Roosevelt, our twenty-sixth president.

"I think Parker deserved rather more credit than any other one man in the entire campaign. . . . He had the rare good judgment and foresight to see the possibilities of the machine-guns," reported TR, the man the rest of the world was calling the hero of San Juan Hill. "He then, by his own exertions, got it to the front and proved that it could do invaluable work on the field of battle, as much in attack as in defense."

BLOOD PLASMA

The Allies were bleeding out. Just months into World War II, British and other Allied troops on the battlefield and British civilians suffering under the Nazi bombing "blitz" of their cities were perilously low on blood for life-giving transfusions. Blood donations simply could not keep pace.

Desperate for help, Dr. John Beattie, the English surgeon in charge of the United Kingdom's blood supply, looked to America to help. But how?

Contrary to popular belief, blood—to this day—can be saved only for up to six weeks before it starts to degrade. Blood plasma, the yellow liquid that constitutes most of blood, could substitute for whole blood in keeping wounded soldiers alive. Predominantly water, plasma also contains vital elements for the human body: electrolytes to prevent dehydration and stimulate neurons and muscle tissue; immunoglobulins to bolster the immune system; hormones, proteins and minerals, and key factors in clotting blood and speeding recovery from burn wounds.

Yet plasma was even more perishable than whole blood, surviving just five days without refrigeration. This was not even enough time for ships to carry it across the U-boat-infested waters of the North Atlantic. Dr. Beattie nonetheless telegraphed an urgent plea to America: "Secure 5,000 ampules of dried plasma for transfusion"—that is, more than the total existing supply of plasma in the world at the time.

The request went to Dr. John Scudder, a Columbia professor and scion of a family with a long and selfless history of medical missionary work in India. Dr. Scudder immediately recruited the best man he knew for the job, someone Dr. Beattie might have remembered as a bright young student he had taught at McGill University's medical school a few years before: Dr. Charles Drew.

Drew hailed from the poor, black section of Foggy Bottom in the nation's capital. One of five children born to a carpet layer and a schoolteacher, he excelled in high school and was able to win a sports scholarship to Amherst, thanks to his skill in swimming, track, and football. After graduating from Amherst, though, Drew discovered that fellowship money for a talented young black man was hard to come by in Depression America, and he worked for two years teaching biology and coaching sports teams at all-black Morgan College in Baltimore. Finally enrolling at McGill, in

Montreal, he finished second in his medical school class of 127 and won both his MD and a master's in surgery.

After teaching at Howard University and working at the Freedmen's Hospital back in Washington, Drew earned a Rockefeller fellowship to Columbia University. There he became the first African American ever to achieve a PhD at Columbia, with a thesis on "banked blood."

It was a propitious moment for such work. Drawing on the work of several leading scientists in the field as well as his own, Dr. Drew developed a process for reducing blood plasma to a powder form and then reconstituting it with distilled water. In this form it could last much longer—long enough to get across the Atlantic and make it to the battle-fields of Europe. Once in the field, medics toted cans carrying two 400 cc bottles: one with the powdered plasma, one with the water. When they were combined, the reconstituted plasma could last for up to four hours.

Drew's organizational genius matched his scientific know-how. Almost overnight, he transformed what had been little more than a series of lab experiments into an industrial-size program to collect, process, and ship out blood in all its forms. Using New York–area hospitals and the donations of some one hundred thousand American servicemen, "Blood for Britain" and "Plasma for Britain" ended up delivering some 14,500 pints of plasma to the Allies, saving countless lives.

As the United States prepared for war in late 1941, Dr. Drew was recruited by the Red Cross to manage a similar blood drive for America. But he was outraged when, a few months into his tenure, the Red Cross bowed to popular ignorance and announced that it would segregate blood by race—a distinction based on nothing but bigotry and supersti-tion. Drew resigned and devoted the rest of his life to teaching at Howard and serving as chief of surgery at the Freedmen's Hospital.

Tragically, one morning, after spending all night in the operating theater at his hospi-tal, Dr. Drew fell asleep at the wheel while driving with some colleagues to the Tuskegee Institute's annual free clinic. He was just forty-five when he died, but he had already left a legacy of service that few will ever match.

THE ARTIFICIAL PACEMAKER

T he dream of some artificial means to keep our hearts beating is an old one, and doctors and engineers around the world have made important contributions to developing a pacemaker. But it was a couple of inventive American engineers, Earl Bakken and Wilson Greatbatch, who turned the pacemaker into a practical reality in our time.

Challenged by her husband to write a horror story during the chilly, bleak summer of 1816, eighteen-year-old Mary Shelley came up with a novel she titled *Frankenstein, or the Modern Prometheus*. Her theme reflected a Europe that was obsessed with the reanimative powers of electricity and had been ever since that colorful American genius Benjamin Franklin—nicknamed, not so coincidentally, "the Modern Prometheus"— conducted his experiments with lightning back in the 1750s. The Italian scientist Luigi Galvani—as in "galvanize"—even posited that all creatures possessed a natural "animal electricity." His nephew, Giovanni Aldini, attempted to prove this by hooking up a primitive battery to the face and rectum of an executed English murderer, George Forster, and letting fly. Forster's jaw dropped and one eye flew open; he writhed about, kicked his foot, and even punched a fist into the air. But he did not, thankfully, come back to life.

Galvani's theories were soon discredited by another Italian scientist, Alessandro Volta—as in "volt"—but the idea of stimulating matter with electricity proved to have as

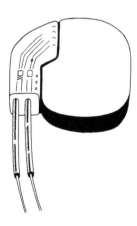

A modern pacemaker, to be implanted in a patient's chest between the skin and the ribs. The wires are attached to the heart muscle.

much life as Mrs. Shelley's novel. By the late nineteenth century, a young British physiologist named John Alexander MacWilliam was speculating that electrical impulses could control the beating of the heart.

An early external pacemaker was invented in Australia. But the term *artificial pacemaker* was coined by a New York heart specialist named Albert Hyman, who, with the help of his brother, Charlie, developed a spring-wound, hand-cranked motor designed to revive stopped hearts—more of a mechanical defibrillator than what we think of today as a pacemaker. (Pacemakers regulate the heart and adjust its rate of beating if it is too fast or too slow. Defibrillators, or ICDs, shock the heart back into beating if it has stopped.) Hyman did not patent or publicize his invention, as such efforts were then characterized as "playing God" by the newspapers.

By the end of World War II, enough hearts had been stopped for a little God-playing to be judged in order. Work on transcutaneous (through the skin) pacemakers accelerated around the world, but most of this involved things like vacuum tubes, large and immobile machines, and wall socket connections that came with a not insignificant risk of electrocution.

Earl Bakken of Minneapolis had a better idea. Inspired by, yes, the 1931 movie version of *Frankenstein*, Bakken at the age of eight devised a five-foot robot of sorts that could talk, blink its eyes, smoke a hand-rolled cigarette, and wield a knife. He also invented a sort of primitive taser, purportedly to keep bullies away. Instructed by a minister to "use science to benefit humankind, not for destructive purposes," Bakken would later build the first wearable external pacemaker for Dr. C. Walton Lillehei in 1958, after Lillehei lost a patient thanks to a local blackout.

Now Dr. Lillehei equipped his heart patients almost immediately with Bakken's artificial pacemaker, which utilized the new silicon transistors. This device was kept in a small plastic box and came equipped with controls to change the heart rate and voltage. Connected to leads that passed through the patient's skin, and to electrodes attached to the surface of the heart's myocardium, it was a great step forward. But it still meant attachments that could serve as a highway for infection from outside the body.

At nearly the same moment, Swedish doctors installed the first fully implantable pacemaker. But problems remained, such as how to recharge its batteries. (Originally an induction coil was used, which must have been little better than what was done to poor Forster.) Many individuals and companies labored to make a better battery over the next decade, but none outdid a Buffalo Sunday school teacher and member of his church choir named Wilson Greatbatch. After getting his engineering degree from Cornell on the GI Bill, Greatbatch invented first a mercury pacemaker battery, then a lithium-iodide cell battery that became the industry standard.

Neither the iodine nor the poly-2-vinyl pyridine in his cathode conducted electricity. But Greatbatch found that when he heated them to 150 degrees Celsius (302 degrees Fahrenheit) they formed an electrically conductive viscous liquid, which maintained its conductivity after being poured into a cell to harden with the lithium. Encased in a titanium cell, Greatbatch's battery proved to have the endurance (energy density), small size, low self-discharge, and reliability needed. Installed, usually, under the clavicle, it could be attached to one or two chambers of the heart, the atrium, and/or the ventricle. Best of all, its life span was soon extended to an average of ten years.

First installed in 1960, Greatbatch's pacemakers were the industry standard by 1971. Medical science continues to make great strides in pacemakers. Today's pacemakers can be implanted with a simple operation, sometimes one that requires only local anesthesia and can be completed in as little as thirty minutes. Leads are fed into the heart via a large vein, with the use of a fluoroscope. When the batteries need to be changed, they can be replaced with another simple operation—one that does not usually need to disturb the ventricular leads. One advance expected imminently is a pacemaker the size and shape of a multivitamin that can be implanted in your leg and control your heart rate from there. Dr. Frankenstein would be pleased.

THE GENIUS DETAILS

Swedish surgeon Åke Senning implanted a pacemaker designed by Rune Elmquist, a Swedish inventor, into Arne Larsson in 1958. It was the first internally implanted artificial pacemaker. It failed after just three hours. Larsson would receive a total of twenty-six pacemakers over the remaining forty-three years of his life and would outlive both the inventor of his pacemaker and his surgeon.

A firewall has now been developed that prevents outside parties from reading the medical information from one's pacemaker or altering it in any way.

The invention of the transistor at Bell Labs in 1947 (see page 81) made possible the modern pacemaker.

Internal pacemakers with mercury batteries, devised by Wilson Greatbatch, were successfully implanted in 1960 at a Veterans Administration hospital in Buffalo by doctors William Chardack and Andrew Gage.

There are about three million people worldwide who now have pacemakers, with another 600,000 receiving implants every year, including roughly 200,000 Americans.

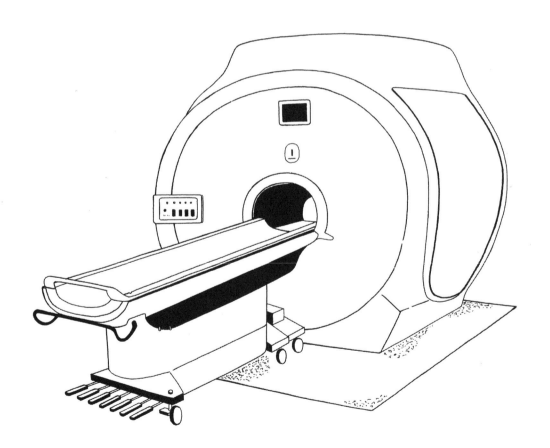

MAPPING THE BODY
THE MRI

T he contributions of immigrants stand out on every page of the long American record of invention, innovation, and achievement. But in no field is the input of new Americans and their children more conspicuous than the development of that miraculous lifesaving device, the magnetic resonance imaging (MRI) machine.

An MRI is not an X-ray, a CAT scan, or a photograph but a radio-created image of the human body that is translated into a picture by computer technology. The idea of magnetic resonance imaging—or "nuclear magnetic resonance imaging" (NMR)—originated with Isidor Isaac Rabi, born in a small town in Galicia and brought to the United States when he was three years old. Small and sickly, surviving—barely—in the slums of New York, he read voraciously, tearing through the shelves of the Brownsville, Brooklyn, public library, attracted especially to the writings of Jack London: "What appealed to me was the democratic idea that anybody could become anything."

Rabi would work his way through a graduate degree in chemistry at Cornell, become the first Jewish professor in the sciences at Columbia, and build the best physics department in the world there. His great strengths lay in research and team building, leading to a breakthrough that determined how a bombardment of electromagnetic waves could reveal the chemical composition of nuclei.

Rabi and his assistants achieved this by first exposing protons to an oscillating magnetic field, which made them line up facing north or south, much as a needle does on a compass. Hit with electromagnetic waves of a precisely calculated frequency, the protons then flipped over. After a fraction of a second, the nuclei relaxed and flipped back, but as they did they "resonated"—that is, they sent out radio signals at the same frequency they received.

All this would mean little—save for the fact that the protons' relaxation period varied according to their makeup. "We Are All Radio Stations" was how a *New York Post* headline summed up Rabi's work in 1939—which was essentially correct: our atoms "broadcast" back what is inside us.

Rabi's discovery would have enormous implications in any number of fields. Scientists used nuclear imaging for all sorts of purposes. But no one thought to use it for purposes of medical diagnosis until Raymond Vahan Damadian, the son of

A modern MRI machine, the product of groundbreaking research by generations of American immigrants.

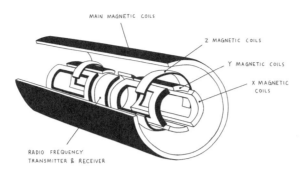

MAIN MAGNETIC COILS

Z MAGNETIC COILS

Y MAGNETIC COILS

X MAGNETIC COILS

RADIO FREQUENCY TRANSMITTER & RECEIVER

Playing you like a radio: a Golay coil, complete with the main magnetic coil that creates a total magnetic field around your body; the x, y, and z magnetic coils that create varying fields to "read" your cells; and the radio receiver and transmitter that receive and convey information from your body.

immigrants who had fled the Turkish genocide against the Armenian people. Born in Manhattan and raised in Forest Hills, young Raymond was a prodigy in several fields, studying the violin at Juilliard's School of Music by the time he was eight, earning money as a tennis pro in the Hamptons by age fourteen, and being selected for a special Ford Foundation college scholarship when he was just fifteen.

Damadian headed for the University of Wisconsin, then the Albert Einstein College of Medicine. He became an emergency room doctor, then a medical researcher, dreaming of curing the cancer that had killed his beloved maternal grandmother.

Damadian was studying how the kidney balances fluids and electrolytes when he audited a Harvard course in quantum physics taught by Edward Purcell, another of the Nobel-winning geniuses who had advanced NMR spectroscopy. Damadian's supple mind grasped a greater possibility before anyone else's did: maybe it was possible to distinguish *cancer* cells from normal ones by means of magnetic resonance imaging. With the human body made up mostly of water, it would be relatively easy to resonate the single proton in hydrogen atoms.

An experiment with the cancerous tumor tissue of rats soon bore him out. Exultant, Damadian proposed a giant leap forward in medical diagnosis: machines big enough to scan a living human being. The NMR machines he was using at the time were built to handle no more than a spinning test tube or a lab slide.

"What I was talking about was like going from a paper glider that you tossed across the classroom to a 747," he admitted. But, "Once you get a strong idea, you can become its prisoner."

Damadian would endure a stiff sentence as the prisoner of his idea, but he could not give it up. Despised as a mere MD by physicists and other PhDs, he was shunned by the cancer research community, labeled a crackpot, and refused funding after his first few years of research. His only backers were his extended family. His wife's brother raised $10,000 on Long Island in the 1974 equivalent of a Kickstarter campaign, even passing the hat at flea markets.

Damadian had to build his own MRI machine with his two assistants, Michael Goldsmith and Lawrence Minkoff. This entailed taking a course at the RCA Institute of Electronics and borrowing software from the Brookhaven National Laboratory (an institution started by Izzy Rabi). Damadian bought thirty miles of superconducting wire, cheap, and he and his assistants spent another year winding it around homemade spools made from metal bookcases, welding an insulated container for the helium the machine would need, and building a radio coil out of discarded cardboard and copper foil tape.

The radio coil had to be wrapped around the body of a very reluctant Minkoff (an experimental mouse had accidentally been "cooked" by the device), the only person on Damadian's team who was thin enough for it to fit. The process of scanning his chest took nearly five hours, until 4:45 on the morning of July 3, 1977, and left Minkoff freezing cold. But when the results were charted, then fed into a computer, they produced the first MRI in history.

Even then, Damadian's troubles didn't end. The National Cancer Institute expressed little interest in the MRI's still relatively crude images as compared to existing CAT scan technology—a bizarre choice, since, unlike CAT scans, MRIs involved no cancer-causing radiation. A friend set up a meeting with General Electric, which Damadian attended, fearful they would steal his idea. Steal it they did, setting off twenty years of litigation.

In 2003, the Nobel Prize Committee awarded prizes for developing the MRI to America's Paul C. Lauterbur and England's Sir Peter Mansfield, whose own MRI machines raced ahead of Damadian's in producing clearer images. Damadian was ignored, even though the written record of their work clearly documented that it originated from his, and even though he held the first-ever patent for an MRI.

Damadian would, in the end, win a boatload of scientific and technological prizes—and $128.7 million, when the Supreme Court ruled that GE had indeed violated his patent rights. His company would go on inventing life-saving devices, such as a mobile MRI scanner and an "open" MRI—work one cannot put a price on.

THE GENIUS DETAILS

The three gradient magnets of an MRI can produce an image of the human body from every possible angle, providing a huge advantage over all other imaging technologies.

Within a year of his first scan of Lawrence Minkoff, Damadian had reduced the scan time of a subject from nearly five hours to thirty-five minutes and demonstrated the difference between healthy and cancerous tissue.

There are now over twenty-five thousand MRI scanners worldwide.

Examined in an MRI machine near the end of his life in 1988, Dr. Rabi remarked, "I saw myself in that machine. I never thought my work would come to this."

The Nobel Prize committee never gave a reason for why Damadian was not included in the prize. Nobel Prizes can honor up to three individuals. The committee's deliberations will not be open to the public for another thirty-eight years.

THE MODERN PROSTHESIS

Van Phillips was a twenty-one-year-old student at Arizona State in 1976 when a motor-boat propeller cut off his left leg below the knee in a water-skiing accident. Back in school on crutches within a week, Phillips soon received another shock: his artificial limb, which he described as "a leg that felt like a fencepost with a bowling ball on the end of it." For the highly athletic Phillips, who had been a pole vaulter, springboard diver, and karate enthusiast before the accident, it was a "most brutal piece of reality."

"I couldn't run on the beach. I couldn't step off a curb. I would hit a pebble no bigger than a dime and fall over it," Phillips said of his prosthesis. "I hated it. I would throw it across the room."

For all the wars of the last century, for all the amputations necessitated as people lived longer and became more prone to diabetes, prosthetics were much the same as they'd been since time immemorial. For legs, that meant a wooden foot on a metal leg, secured by straps to a leather knee (or hip) casing—a state of affairs reduced to the acronym SACH, for Solid Ankle Cushion Heel, about the best that was thought could possibly be done for the amputee. The whole device was lifeless, a sort of stand that sent a shock back up the spine every time its wearer landed on the fake heel.

Phillips had no medical background, but he decided to build a better leg himself. That first required years of education in biomedical engineering, but he never stopped thinking of how he might regain his old, athletic lifestyle. Practicing tae kwan do, he decided that what he needed was a prosthetic that dispersed shock resistance and gave him at least an inch of bounce; one that could bend and fit the body as it changed, without the usual adjustments that had to be made in prosthetics when their users lost and gained even small amounts of weight.

Hiking for hours, pulling his unwieldy false leg with him, Phillips looked everywhere in nature for some model. Then it came to him: the cheetah, the fastest animal on earth. The key to its speed was how the long tendon of the big cat's hind legs curved in the shape of a "C" from hip to foot. Every time it hit the ground, that tendon stretched like a catapult, firing the cheetah forward again. Its foot stored energy every time it landed, then fired it back out again.

Phillips began to search for the right material, going through titanium, plastics, steel chrome, rubber, and aluminum before deciding on a light carbon graphite foot. To his delight, it allowed him to walk better and even run and play tennis. But he still found himself breaking a foot a week, something that plunged him into depression each time he had to mold a new one.

Then Phillips met Dale Abildskov, a specialist with an aerospace company, who moved him away from some imitation of a foot and toward making the entire prosthesis like that C-shaped tendon. Phillips knew the moment he tested it that the idea was sound, transferring more energy to and from the "blade" of the prosthesis with each step forward than the human foot could ever provide. It didn't even need a heel.

Phillips quit his job the next day and set about perfecting the new leg. It only took . . . about two years, as he went about building—and breaking—some two hundred to three hundred more artificial "feet." Working in his basement laboratory, he constantly risked serious injury and nearly sustained a fatal accident when he inhaled some toxic fumes that laid him up for a week, unable to eat, speak, or move.

Van Phillips found that only carbon graphite could compete with human or animal tendons and ligaments for strength and agility. Each individual strand of carbon graphite is thinner than a human hair, but tens of thousands of these strands, bound together with a gluelike carbon epoxy matrix, can make a material that surpasses even the capacity of human tendons to store and release energy.

Abildskov not only helped Phillips to reengineer his foot but also guided him toward obtained financing and support. Phillips, Abildskov, Bob Barisford (Abildskov's old boss at Fiber Science, his aerospace company), and another partner, Walt Jones, joined them to form Flex-Foot in 1983. Bob Fosberg would later bring his expertise as a Harvard MBA and pharmaceutical executive to the new company.

"There's nothing as much fun as getting four or five people together, all with the same goal," Phillips would later tell groups of young people about their efforts. "Each of us would spur on the others, and you get this vortex of energy as the idea goes round and round."

Van Phillips's cheetah-inspired prosthetic leg. The blades are almost five times more efficient than the human ankle in retaining energy with each thrust.

Van Phillips holds sixty patents today for designing prosthetic legs that allow him to run and walk, and also ones specifically for skiing, swimming, surfing, scuba diving—and riding horseback, with his daughter.

Each prosthetic "Cheetah leg" costs between $15,000 and $18,000 today.

The Cheetah blades put runners in a competition at a disadvantage in that they do not allow thrust at the beginning of a race. Runners using Cheetah blades must generate more power from their gluteal and abdominal muscles.

Some studies have shown that after the runner has reached maximum speed the blades do offer a theoretical advantage, as they lose only 9.3 percent of energy while hitting the ground, while the human ankle loses 42.4 percent of energy.

What they produced was a stunning breakthrough in the prosthetics world, and while Phillips sold the venture to the Icelandic firm Ossur in 2000, his work nonetheless became famous as the design that allowed Irish American athlete, model, and actress Aimee Mullins, who lost both her legs below the knee before the age of one, to become a Paralympic champion. An estimated 90 percent of all Paralympic athletes today use a Phillips-designed prosthetic. His blades made it into the 2012 Olympics, and many suspect that one day the fastest athletes in the world will be amputees fitted with "superior" new legs. Some even call it part of an evolution to a new form of human being, while others are looking forward to the continuing advances in myoelectric prostheses that promise to allow people to control artificial limbs by their minds, and even to regain feeling in them.

Phillips himself predicts that the ultimate solution will be "regenerated limbs," but he's not waiting for that. Instead, ever since selling his company, he has devoted his energies—and over a million dollars of his own money—to developing advanced prosthetics that will cost no more than ten dollars and be easily repairable for the estimated five to ten million individuals who have lost legs and feet to the land mines that now litter the developing world. Will he succeed? As Van Phillips likes to tell classes of children: "Anything you can think of, you can create."

HEAL THYSELF
GENE THERAPY

T he prospects were grim in 1990 for four-year-old Ashanti DeSilva, who suffered from a rare ailment known as adenosine deaminase (ADA) deficiency, which caused severe combined immunodeficiency (SCID), leaving her prey to infections or viruses that could easily kill her. The problem was a single defective gene that kept Ashanti's cells from producing the enzyme that makes white blood cells what they are: the body's natural protectors against the myriad diseases and viruses that routinely attack us.

But doctors hoped to help DeSilva with the new tool of gene therapy. This had been anticipated for many decades. Experiments in 1952 had confirmed DNA (deoxyribonucleic acid) as the molecule of genetic information. Then in 1953 James Watson and Francis Crick discovered its double-helix structure. By 1966, the genetic code was largely deciphered, thanks to the work of Marshall Nirenberg and three of his colleagues at the National Institutes of Health (NIH), as well as other scientists around the world. Only a few years later, in 1972, Drs. Theodore Friedmann and Richard Roblin proposed that undamaged DNA could replace defective or ineffectual DNA in patients suffering from hereditary disorders, in a process that would become known as recombinant DNA technology.

The basic idea of Ashanti's gene therapy was to draw her blood, treat her defective cells with the working gene she lacked, then inject the healthy cells back into her. The result: the patient lived—but the procedure was far from a complete success. The treated cells managed to produce the enzyme, but they did *not* produce more healthy cells, thereby sinking hopes—for the time being—that the body could be trained to repair itself indefinitely.

DeSilva remains alive and well as of this writing, but she still periodically repeats her gene therapy treatments and takes a drug that contains the needed enzyme. Her treatment made it clear that gene therapy was more complicated than had been hoped. In 1999, the terrible death of Jesse Gelsinger, an eighteen-year-old who had volunteered to take part in gene therapy trials for an inherited liver disease that he had been able to control with drugs, sparked a crisis of doubt in (and about) the field. Gene therapy projects around the country were suspended or canceled, funding was slashed, lawsuits were launched, ethics panels were convened, and the entire treatment fell into ill repute.

The problem lay both in the system for delivering those repaired, healthy genes via "vectors"—"deactivated" or "hollowed-out" viruses—and in the sheer complexity of the human body. A previous, maybe unsuspected exposure to the virus could set off

Philip Leder, a geneticist working with Nirenberg, developed a method for reading the genetic code on pieces of transfer RNA in 1964, thereby accelerating the deciphering of the genetic code.

The Human Genome Project, which has plotted the entire sequence of genes in the human body, estimates that each human being has twenty thousand to twenty-five thousand genes.

Glybera, a gene therapy product for treating a protein production deficiency, costs $1.6 million per patient and is estimated to be the world's most expensive drug.

Somatic gene therapy—the repair, transformation, or replacement of a person's individual genes—is allowed in the United States, as it affects just the individual patient, not his or her offspring.

Children born with donated ooplasm—a key part of an embryo's formative yolk—to mothers whose own ooplasm is defective have three genetic parents.

an extreme counterattack by the body's immune system (possibly what proved fatal for Jesse Gelsinger). Delivery to some of the wrong genes could cause them to stop functioning and allow the rise of cancerous tumors—something that victimized other gene therapy subjects.

Yet the research effort in labs throughout America has recovered, and dozens of trials are now reaching their late stages, promising ways to deliver genes with safer and more effective viruses. There have already been some encouraging early results in combating more immunological disorders, as well as hemophilia, anemia, and muscular dystrophy; neurodegenerative diseases such as Parkinson's and Huntington's; viral infections such as HIV, hepatitis, and influenza; heart disease, diabetes, numerous cancers, and even blindness. Researchers have traced some four thousand diseases to defective genes, from amyotrophic lateral sclerosis (Lou Gehrig's disease) to Alzheimer's to arthritis. The delivery vectors, or viruses, they are working with may soon be able to repair, replace, or even "silence" genes gone awry (those that create cancers or kill brain cells), reaching every part of the body and crossing the blood-brain barrier. Eventually, "improved" genes may even be able to do things like kill your urge to smoke—or endow your offspring with advanced abilities and powers in any number of endeavors.

This potential raises obvious ethical questions about just where "gene therapy" becomes "genetic engineering," a much more dubious concept. It has resulted in the banning of "germline," or hereditary, gene therapy, at least until we can better understand all the consequences. For the time being, gene therapy also remains almost prohibitively expensive. But as of 2014, some two thousand gene therapy clinical trials around the world have been conducted or approved. Over six hundred such trials are under way in the United States. As our science continues to advance, and as US and foreign companies continue to pour money into gene therapy research, it seems inevitable that one day we will be able to cure some of what are now our most formidable diseases with a quick injection—and maybe raise our future progeny's SAT scores at the same time.

MR. WHITNEY'S MACHINE
THE COTTON GIN

I t is the supreme irony of US history that the South's agrarian, antebellum slave economy was rescued by a mechanical device—invented by a Yankee businessman.

At the start of the nineteenth century, the South was a region in search of a purpose. Overwhelmingly agrarian, its population included over seven hundred thousand enslaved African Americans, growing tobacco, rice, and indigo—all waning industries. Europe, the biggest and richest market in existence, could find these commodities and many more at lower prices elsewhere in the world.

There was always cotton, which grew so abundantly south of the Mason-Dixon Line that Alexander Hamilton wrote, "Several of these Southern colonies might some day clothe the whole continent" with it. Cotton was already an enormously important product in world trade, feeding the bustling cloth mills of England that were at the forefront of the Industrial Revolution—but nearly all of it came from India. Long-staple cotton was farmed profitably along the coasts of Georgia and South Carolina,

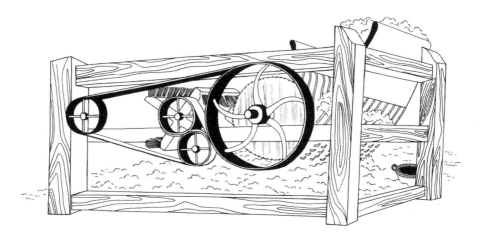

A plantation-sized cotton gin, near the beginning of the American cotton boom.

its black seeds easily removed by rolling presses, but it would not grow farther inland. Superior, short-staple cotton required a person working fourteen hours a day to pull all the green seeds from the short, dense fibers of just one pound of the stuff. Cotton, it seemed, was too labor-intensive even for a slave economy.

Eli Whitney to the rescue. Born in central Massachusetts in 1765, young Eli started his own business at age eleven, crafting nails—a hardware badly lacking during the Revolution—at his father's forge and personally selling them to outlying farms. When the war ended and nails again proliferated, Whitney noticed a change in women's fashion and turned to making hatpins. Soon he was selling walking canes as well. He earned his own tuition to Yale by teaching in local schools, and graduated with a law degree, but finding few opportunities headed for Georgia to take a tutoring job.

Along the way he fell in love with Catharine Littlefield Greene, the mistress of Mulberry Grove. That was all right; everyone did. "Caty" Greene was herself a transplanted Yankee and the belle of the Revolution. Married to Washington's most capable commander, Nathanael Greene, Caty bore him five children, while following the army and even sticking it out for the winter at Valley Forge. Hamilton, Thomas Jefferson, even Washington himself were all smitten by her extraordinary vivacity.

Mulberry Grove was a plantation, a gift from the state of Georgia to General Greene, partial payment for the huge debts he'd incurred financing his own troops in the field, but the native Rhode Islander soon died of sunstroke. Caty Greene and Phineas Miller, the supervisor—the friend of Whitney's who had set up the tutoring job for him—kept the plantation going, but it was hard work. Like many of their neighbors, they were bedeviled by the question of how to make cotton pay.

In a classic male effort to impress, young Whitney busied himself fixing everything he could find around the plantation. Impressed, Greene introduced him to local planters also looking for some way to get seeds out of cotton, telling them, "Gentlemen, apply to my young friend, Mr. Whitney. He can fix anything!"

What Whitney came up with, just nine days later, was the cotton "gin" (short for "engine"). The picked cotton was fed through two contra-rotating drums edged with teeth that pulled the cotton from the seeds. The fibers were then fed through a wire-mesh sieve, leaving the seeds behind, while a fast-rotating brush swept away the lint.

The cotton gin was almost infinitely expandable. In the course of a working day, the amount of cotton that could be processed by a single individual using a hand crank rose from one pound to fifty. Powered by horses or a water mill or, later, a newfangled steam engine, a bigger gin could clean up to one thousand pounds of cotton a day. All it required was one or two people—that is, enslaved people—feeding it through the drums, and another couple to gather the seedless cotton fiber on the other end.

The story would later spread that the design of the cotton gin was really the work of Caty Greene, as women were not allowed to file patents at the time. This is dubious, considering Whitney's demonstrated mechanical skills, but Greene and Miller—now married, alas!—did back Whitney's invention with all their money and connections. His patent was signed by President Washington himself. But cotton planters throughout the South had been working on some sort of gin for years, and Whitney made a fatal error by first trying to charge them one-third of their cotton to use his machine.

They violated his patent at will, engaging him in years of nasty and ruinous court battles. Whitney switched to manufacturing cotton gins for sale, perfecting a better version of his machine that used a fine-toothed circular saw for the teeth. He invented one that even bagged the cotton as well, at his new factory in New Haven, Connecticut. But a series of local fevers and a catastrophic fire delayed production, and by the time the courts finally ruled in his favor in 1807, Whitney was nearly bankrupt. He had, in any case, moved by then into a new line of work: manufacturing guns.

Meanwhile, the share of all the world's cotton grown in the United States rose from 9 percent in 1801 to 68 percent by the Civil War—even though cotton production around the planet had tripled. By 1860, the South was producing 2.275 billion pounds of raw cotton, 60 percent of *all* exports from the United States, supplying over 80 percent of the cotton for clothing manufactured in Britain and every bit of what was used in New England's fast-growing mills.

Perversely, the triumph of mechanization had rendered agricultural slavery economically viable again. Freed from cleaning cotton, more slaves than ever could be put to work planting it, picking it, and clearing the land to grow still more of it. By 1860 there were nearly four million slaves, making up three-eighths of the total population of the South, and a majority of several southern states. Civil war loomed—a tragic consequence of an invention magnificent in its simplicity.

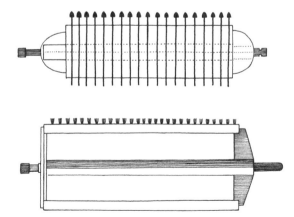

The picked cotton was fed through rotating drums with their pointed edges, or "teeth," to remove the seeds (top), then cranked out through a wire-mesh sieve (bottom), leaving the seeds behind.

MECHANIZED FARMING

How momentous is an invention when it is able to help push a whole country hundreds of miles west—*and* into the future?

Oxen-pushed reapers were used by ancient Celts and Romans before being lost to the Dark Ages. By the early 1800s, after more than a thousand years of grains being harvested mostly by men with scythes, reapers were being reinvented in England and Scotland, but they had become almost a mania in America. The harvest was a time of maximum peril for the farmer, for if it was not to spoil, grain had to be brought in as soon as it ripened, something that often meant paying top dollar to hire unknown drifters to work and live with your family for weeks.

How to come up with something better? The basic design for a "mechanical reaper" was a wooden platform, pulled by horses walking to the side of the wheat. Sticking out from the side, rows of wire fingers grasped the wheat stalks; a revolving wooden flail pulled the stalks to the moving blades, which cut them, then lifted them back to a platform, where they could be swept off by a man with a rake. A main wheel kept the whole contraption running.

By 1833, the most successful such device was the Hussey Reaper, patented by Obed Hussey, a formidable ex-whaler from Maine, but he already had competition. Cyrus McCormick, a Virginia farmer whose father had spent twenty years trying in vain to invent a reaper, entertained the McCormicks' neighbors with a machine that cut six acres of wheat in the time it took six men to do the same with scythes.

McCormick's small (but crucial) innovation was to place a triangular metal wedge at the front of the reaper in order to separate the wheat being cut from the rest of the crop, and thus keep the field from looking like a bad punk haircut. Setting out to crush Hussey, McCormick and his family lost what came to be called "the War of the Reapers" in patent court but won in the fields. A large, grim, imposing individual with a massive head and beard, McCormick was "a great commercial Thor," as one biographer called him. He personally challenged Hussey's and others' reapers to public grain-cutting competitions, which he won easily. Seemingly devoid of any other interests save church, McCormick worked fourteen hours a day, squeezing every last dime of extra expense out of his business.

Like many early American business titans, he possessed an almost innate feel for self-promotion, regularly touring Europe with his "mechanical man," winning prizes and gold medals, and building a worldwide market for his product. At home, McCormick was an early believer in advertising and especially credit. He asked only $30 down for reapers that cost $120 to $130, accepting the rest on time and charging only the interest the banks charged him. If his machines did not cut at least one and a half acres in an hour, customers could have their money back: "One Price to All and Satisfaction Guaranteed."

In a precursor to auto dealerships, agents were hired across the West to teach McCormick clients how to use their new machines, forging bonds that would last across generations. With the aid of his father and brothers, McCormick improved his reapers every season and bought up patents and companies from his many competitors. He was even able to acquire Hussey's superior, sawlike cutting bar in 1850.

Yet his best move was west. McCormick was amazed by the size of the farms he encountered on the gently rolling prairies of Illinois—lands perfectly suited to his machines. He was perceptive enough to see that a ramshackle little mudhole of a town called "Chicago"—invented the same year as his machine—was ideally situated to become the new headquarters of his enterprise, but he had only $300 in capital on him. Problem solved by the settlement's indefatigable booster, William Ogden (see page 155), who offered him $50,000, cash, in exchange for a half interest in his reapers.

McCormick accepted, and built a monolith for the time: a 40-by-100-foot, two-story factory, run by a 30-horsepower steam engine. By the end of his first year, he had made and sold 500 reapers. By the end of his second year, he'd made 1,500 more and was able to offer Ogden a 100 percent profit to buy back his half interest. Ogden cheerfully accepted, and not simply out of civic duty. McCormick's reapers meant outbound freight for the rail lines Ogden was developing throughout the Midwest and all the way out to the Pacific. They meant countless golden bushels of grain going out, and grain-fed livestock coming back.

Above all, McCormick, twenty years before the Civil War, had moved his business out of the soil-draining, soul-breaking cotton and tobacco agriculture of the South and into "the broad, sunlit uplands" of the American future, where capital might be had and grains might be produced on a staggering scale. He didn't do it alone. A bevy of other agricultural inventors pushed McCormick to make his own farm implements better and better—and were frequently bought out by him.

Jearum Atkins came up with a "self-rake" that would collect cut grain from the reaper platform. Charles W. and William Marsh added a harvester on which two men could stand at a table and bind the grain as it came up. Charles Withington and John Appleby invented mechanical binders that would do the same work automatically. Hiram Moore was the first to put it all together in huge "combine harvesters" pulled by sixteen or more horses, and by 1911 Holt Manufacturing had steam-powered mechanical combines harvesting everything from wheat to bananas. What had been weeks of backbreaking toil

The "War of the Reapers" ended tragically in 1860 when the sixty-eight-year-old Obed Hussey, fetching water for a thirsty little girl, fell under a moving train.

In 1830, it took four people and two oxen ten hours to harvest two hundred bushels of wheat. By 1895, six people and thirty-six horses, working combine harvesters, could produce twenty thousand bushels of wheat in ten hours.

American wheat production tripled from 1869 to 1919, from 250 million to 750 million bushels. The United States exported 200 million bushels of wheat to war-torn, starving Europe between 1914 and 1922.

American wheat production peaked at 1.228 billion bushels a year in 1945–48. The United States sold 150 million to 440 million bushels a year to the USSR between 1963 and 1980.

McCormick's descendants would build a newspaper empire around the United States, including the first American tabloid, the *New York Daily News*.

by two dozen men became a few hours' work by a single driver in an air-conditioned cabin.

McCormick's company would merge with several others into the global giant International Harvester. Along with the likes of Allis-Chalmers, John Deere, Gleaner Manufacturing, Minneapolis-Moline, and the Ford company's trucks and tractors, they would increase exponentially, over and over again, the yield of American fields, at a tiny fraction of the cost and manpower. Their machines and methods would spread around the world, used and emulated even by the likes of the new communist regime in the Soviet Union. America would become the breadbasket of the world, able to export half its wheat to other countries.

THE SUPERMARKET

I t all started thanks to an iceberg.

John Jacob Astor IV made the mistake of returning from England in 1912 aboard a spanking new ocean liner called the *Titanic*. When the fateful collision came, he helped his pregnant wife to a lifeboat, retired to the rail to smoke a cigarette, and was never seen alive again. Most of his $87 million real estate empire was left to his son, Vincent, a student at Harvard who had never shown much interest in anything but fast cars and model farming.

Taking the reins, Vincent embarked on an extraordinary array of rather beautiful building projects around Manhattan, including a twenty-one-thousand-square-foot, one-story market arcade at Ninety-Fifth Street and Broadway, completed in 1915. Made out of mottled travertine marble and festooned with banners, the market offered something new: stands where farmers and merchants sold all sorts of fresh meat, fish, produce, and other foodstuffs in one convenient location. It was perhaps—depending on how you see it—the world's first supermarket.

Previously, shopping in America—and everywhere—had consisted mostly of going into countless individual shops and handing sales clerks lists of what you wanted. They then proceeded to fetch, measure, cut, and wrap them while you gossiped with your friends or, presumably, fell into a fugue state. A few stores, most notably the Great Atlantic & Pacific Tea Company, better known as the A&P, a former mail-order tea business, had grown into formidable chains by selling any number of "dry goods"—canned goods, coffee and tea, staples such as sugar or flour. But no one thought it a very good idea to add more perishable items, what with there then being one corner grocery for every four hundred Americans. So it proved with the Astor Market, which was out of business in two years, replaced by a skating rink.

Clarence Saunders thought this was wrong. A poor boy from Virginia, Saunders had left school at fourteen and bounced around the South for years, working at a coke plant in Alabama and a sawmill in Tennessee before moving to Memphis and organizing a wholesale grocery cooperative when he was twenty-one. There he became convinced that the biggest problems for most groceries were high overhead and bad credit losses.

By 1916, when he was still just thirty-five, Saunders had accumulated enough capital to put a better plan into action. He opened the world's very first single-owner, self-service grocery store. After passing through a turnstile, customers were provided with a shopping basket and allowed to roam through the aisles on their own, picking out what they wanted before paying for it at cash registers near the front. No need for a legion of clerks to fetch items. Customers could now see full displays of all the store had to offer, creating a critical new revenue stream: impulse buys. At the checkout, it was cash only—no talking your neighborhood grocer into carrying you another week.

Saunders quickly patented "the Self-Serving Store," which he named "Piggly Wiggly." (Asked why, he would usually reply, "So you would ask me that.") It looked more like a stockyard than a supermarket, with low railings and chain-link fencing, but it was the right store at the right time. America was about to enter World War I, which would send food prices soaring and enhance Saunders's advantages of scale and savings on labor. A mastermind at marketing, he kept improving and publicizing his stores, with their unique pig-pink decor and low prices.

Piggly Wiggly fattened up with astonishing speed. Within six years, Saunders had 1,200 stores in twenty-nine states, Piggly Wiggly was listed on the New York Stock Exchange, and the poor-boy-made-good was building himself an only mildly grandiose mansion, from porcine Georgian marble, called "the Pink Palace."

He would never live in it. Inclined to get ahead of himself, Saunders got embroiled in a battle over company stock with Wall Street short-sellers, and lost. He also picked a bitter, personalized fight with the all-powerful political boss of Memphis, E. H. Crump. Before the 1920s were out, Piggly Wiggly was off the stock exchange and Saunders was gone from his own company, out $3 million and forced into bankruptcy. The Pink Palace was sold off to the city of Memphis.

Saunders tried to recoup by opening a new chain in 1928, the "Clarence Saunders Sole Owner of My Name" stores. Even with that moniker, Saunders's new enterprise flourished, reaching 675 "Sole Owner" stores and $60 million in sales within a year. A football team he sponsored for promotional reasons—yes, the "Clarence Saunders Sole Owner of My Name Tigers"—thrashed the Green Bay Packers and received an invitation to join the NFL—which, of course, he declined.

The Great Depression soon put paid to Sole Owner, but Saunders moved on to his next idea, a sort of supermarket automat or giant vending machine that he called "Keedoozle," for "Key Does All." All products were displayed in glass cases. Customers were handed a key on entering the store, which they plugged into a slot beneath the product they wanted. They then pulled a trigger on the key to indicate how many of each item they wanted. This would be recorded on a punch tape, which activated back-office machinery, leading to the chosen goods being assembled and put on a conveyor belt to the cashier, where the bill was totaled by a machine reading the punch tape.

If you're asking yourself what could possibly go wrong with such a system and answering, "Everything," you will not be surprised to learn that Keedoozle was soon keeput. Still undaunted, Saunders was planning yet another automatic store at the time of his death in 1953, a forerunner of self-checkout in which the customer, using an early sort of hand-held computer, or "shopping brain," would "act as her own cashier."

"I can handle a $2 million volume with only eight employees," he calculated.

We will never know. But there is no doubt that Saunders's Piggly Wiggly inspired the inception and expansion of countless supermarket chains across the country. The A&P began selling perishables and creating its own brands soon after World War I. From 1915 to 1975, it was the largest food or grocery retailer in the United States; from 1915 to 1965, it was the largest retailer in the United States, *period*.

By the 1960s, the competition had sparked all sorts of innovations to draw customers. Patrons earned discounts and gifts by collecting "green stamps"—think club cards—and stores routinely gave away, week by week, entire sets of children's encyclopedias and histories of the world or the United States. (The author was the proud owner of three different sets by the age of eight.) As America became more and more of an affluent, car-oriented, suburban nation after World War II, individual supermarkets became larger and larger, carrying more and more products of all kinds, including medicines and toiletries and prepared foods. They also added conveyor belts of steel rollers, with the bagged groceries placed in boxes and slid right out of the store, to be collected at curbside by the driver at her leisure.

Clarence Saunders would have been proud.

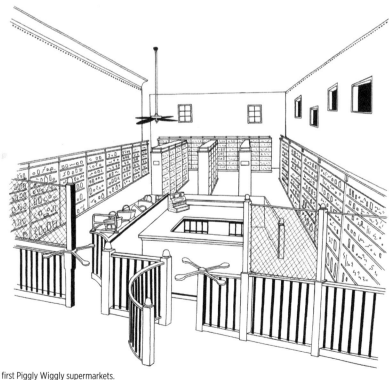

The layout of the first Piggly Wiggly supermarkets.

INVENTING A CITY: L.A.
SOUTHERN CALIFORNIA AQUEDUCT

W hen El Pueblo de Nuestra Señora la Reina de los Ángeles passed into the hands of the United States at the end of the Mexican War, it was an agricultural community of 1,610 souls who eked out a living on land irrigated from the shallow, meandering Los Angeles River. Little more was thought to be possible. Unlike San Francisco, Los Angeles had no harbor and no nearby goldfields, and it averaged only fifteen inches of rain a year.

It started to grow after the Civil War, thanks mostly to an ambitious chamber of commerce and local sanitarium doctors who made L.A. "the best advertised city in the country." By 1900, Los Angeles was a bustling city of over one hundred thousand, but the surrounding desert had not gone away. The lack of water obsessed William Mulholland. An itinerant Irish immigrant, Mulholland had come to California in 1877 as a common sailor, walking across the Isthmus of Panama because he could not afford the train fare. Unimpressed by L.A., he started for the port of San Pedro to take to the sea again but was offered a well-digging job along the way. Soon he was cleaning the open ditches and log pipes of the privately owned Los Angeles Water Company. Within eight years he was superintendent.

"The city is condemned to grow," he liked to say—but only if there was enough water. Mulholland instituted water meters in 1889, far ahead of his time. They cut per capita consumption by a third—but the city kept growing.

"If you don't get the water, you won't need it," Mulholland warned, with Yogi Berra–like pith.

Fred Eaton, Mulholland's former boss, who went on to become mayor of Los Angeles, made the water company a city department. A camping trip with Joseph B. Lippincott, regional engineer of the US Bureau of Reclamation, gave Eaton his first inkling of where more water might come from: Owens Valley, a bucolic farming and ranching community nestled under the slopes of the Sierra Nevada, 225 miles away.

The Los Angeles Aqueduct today, bringing water 225 miles to the city.

The aqueduct originally had six storage reservoirs. Altogether, including unlined and open canal, the water traveled 233 miles.

By the time the aqueduct opened, Los Angeles was already drawing people from around the world, including 250,000 Mexican Americans, 40,000 Japanese Americans, and close to 50,000 African Americans, most of them from the South.

Workers on the aqueduct were paid $2.25 a day. They lived in fifty-seven camps along the construction sites, in sleeping tents and bunks. A total of forty-three workers died on the job because of accidents.

A 2006 ceremony at the same site where Mulholland had started the flow of the aqueduct into L.A. diverted some of the water back to the Owens River. David Nahai, president of the L.A. Board of Water and Power Commissioners, said on the occasion, "There it is. . . . Take it *back*."

Today the entire aqueduct system provides 430 million gallons of water to Los Angeles every day.

Eaton hired Lippincott as a city "consultant" on water rights, a de facto bribe. Lippincott then helped Eaton buy up water rights in Owens Valley, telling the locals the purchases were all part of an irrigation system he had promised them.

Their corrupt bargain was backed by some of the most powerful men in California. The Owens Valley farmers and ranchers protested bitterly when they found out what the real deal was, insisting that Los Angeles did not need nearly as much water as the city was claiming but wanted to use the excess to power electrical plants and develop land in the San Fernando Valley. Their congressional representative appealed all the way to President Theodore Roosevelt himself—who found in favor of Los Angeles, ruling that the water "is a hundred or a thousand fold more important to the state and more valuable to the people as a whole if used by the city."

Mulholland assembled a workforce that reached 3,900 men, most of them immigrants from Europe and Mexico, to build a four-thousand-foot aqueduct from the valley to the city. It was an epic construction project. The aqueduct operated by gravity but required the building of five hundred miles of road, the laying of a 120-mile railway, and the stringing of 377 miles of telegraph and telephone wires just to transport men and materials. It also took six million pounds of blasting powder to blow out forty-three miles of tunnel, and the building of three cement plants to construct the aqueduct itself.

Crossing desolate Jawbone Canyon alone required a 3,126-ton, 8,095-foot-long steel siphon with plate as much as an inch thick. There were ninety-eight miles of covered conduit, some of it wide enough to drive a car through, and fully exploiting the water it brought would require the construction of two hydroelectric plants, six storage reservoirs, and 218 miles of power lines. Scheduled to take five years, it was finished twenty months ahead of time.

As the aqueduct's waters swirled into a canal on November 5, 1913, Mulholland told the crowd of forty thousand thrilled Angelenos who turned out to see the tap opened: "There it is. . . . Take it."

His words would become an operating slogan for the city of Los Angeles. By the time the aqueduct was finished, Mulholland had predicted, L.A. would have 260,000 residents. Instead, there were 485,000, then 1 million by 1920, and nearly 2.6 million by 1930. Growing eleven times faster than New York City, it became one of the world's great polyglot cities, drawing immigrants from all over the world and providing rare opportunities for even the poorest of them. Los Angeles also grew from 61 to 440 square miles. Not yet a car city, it remained a remarkably clean, healthy metropolis, run by the electricity all that water power generated.

Meanwhile, the Owens Valley farmers rebelled, launching eleven separate dynamite attacks on Mulholland's aqueduct in what became known as "the Water Wars." But by 1928 the city had won. It owned 90 percent of the water in Owens Valley, and almost all agriculture there was finished. The alkaline residue from its now dry lake bed created some of the worst air pollution in the country.

Fred Eaton, who had been compensated by the city for the $450,000 in water rights he'd bought for L.A. from Owens Valley residents, now demanded $1 million for land he owned in order to build a dam and storage reservoir. Mulholland refused, building more storage reservoirs himself. There were soon signs that one of them, the St. Francis Dam, forty miles north of Los Angeles, was on the verge of collapse. Mulholland went to inspect it on March 12, 1928, and deemed it safe. Hours later, the dam gave way, loosing a seventy-five-foot wall of water over most of Ventura County. An estimated six hundred people were killed, most buried by mud. It was one of the worst peacetime disasters in American history, and Mulholland, accepting full responsibility, immediately resigned. Eaton never did get his price and died bankrupt.

"The water octopus of Los Angeles," as the *Los Angeles Times* called it, continued to grow. Extending its tentacles to the Mono Basin and Feather River, six hundred miles distant, then to a bold new dam the federal government was building on the Colorado River (see page 171), it added a second aqueduct, and more dams and reservoirs, over the twentieth century. But the demands of other states and cities, and California's extended drought in the twenty-first century, would finally pull it back to an emphasis on conservation. Having slipped all predicted bounds in its past, the City of Angels awaits a new sort of ingenuity to help it flourish on a warming planet.

THE PHONOGRAPH

Contrary to what we were taught in grade school, Thomas Alva Edison did *not* invent everything under the sun—just *almost* everything. Before he was finished, the "Wizard of Menlo Park" would hold 1,093 US patents—an extraordinary accomplishment for a man with no formal education beyond the eighth grade.

His favorite invention, the phonograph, would come easiest of all—and it was the one he almost let slip away. The idea for it stemmed from Tom's teenage days as a tramp telegraph operator, struggling to translate the rapid-fire clicks of Morse code into words at the rate they came in. His solution was to find an old machine on which the dots and dashes were indented on a paper tape as they were received, then to run the tape through a second machine at a speed more conducive to the relationship between human hand and mind.

In December 1877, experimenting with ideas for improving on Bell's new telephone (see page 74), Edison applied this same trick. He hollered "Halloo!" into a telephone diaphragm that had an embossing point attached. The sound moved the point across waxed paper. When the indented paper was moved back through the machine, sure enough, Edison could hear a vague echo of his greeting.

Edison—still just thirty—had a mechanic named John Kruesi construct a machine to replicate this experiment. He then shouted the words to "Mary Had a Little Lamb" into the diaphragm while turning a shaft handle. This caused a stylus to vertically emboss a cast-iron cylinder wrapped in tinfoil. He wound the cylinder back, hoping it might have picked up a word or two, and was amazed to hear a clear recording—the first real recording ever made, by anyone—of his voice.

"Mein Gott in Himmel!" swore Kreusi.

"I was never so taken aback in my life," Edison himself recalled. ". . . I was always afraid of things that worked the first time. Long experience proved that there were great drawbacks found generally before they could be got commercial; but here was something there was no doubt of."

His doubts did not stop him from promoting his latest invention. *Scientific American* reported that just eighteen days later, on December 22, 1877, Edison visited the magazine's offices: "He placed before the editors a small, simple machine about which very few preliminary remarks were offered. The visitor without any ceremony whatever turned the crank, and to the astonishment of all present the machine said: 'Good morning. How do you do? How do you like the phonograph?' The machine thus spoke for itself, and made known the fact that it was the phonograph."

Yet Edison's doubts about the commercial value of his machine were justified. The tinfoil of his phonograph shredded easily. Its cylinders could provide only distorted recordings two minutes in length, and there was no means of mass-producing them. Edison, distracted by his work on the electric light (see page 167), did something he almost never did: sell and license his patent on "the phonograph." This led to little, save for some truly frightening German "talking dolls" that most parents rushed back to the store.

His departure from the field gave some of the rival geniuses at Alexander Graham Bell's Volta Laboratory, down in Georgetown, their opening. First, in 1886, Bell's cousin, Chichester Bell, and Charles Sumner Tainter invented the Graphophone, which engraved sound in a lateral, "zigzag" pattern—instead of Edison's "hill-and-dale" (up and down) vertical grooves. It used a cardboard cylinder covered in wax. In 1887, Volta's Emile Berliner, a high-strung German-Jewish immigrant who had invented a sort of early microphone and who later created a prototype of the helicopter, came up with the idea of recording on a *flat disc*—an idea Edison had previously considered and rejected.

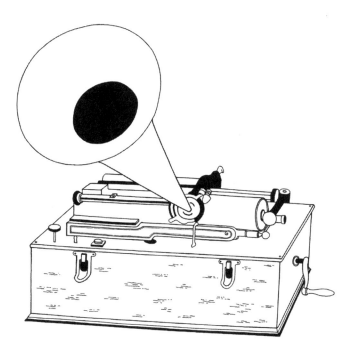

An advanced Edison phonograph,
complete with megaphone.

This was the Gramophone, and it had the great advantage of making mass-produced recordings possible. Berliner's stylus would engrave the sound on a thin coat of wax over a zinc disc. The disc was then etched in chromic acid and electroplated—a process that turned it into a stamper, which could be subsequently pressed into a ball of hard rubber "Vulcanite" to mass-produce copies.

Volta offered to combine its efforts with Edison's, but the Wizard rejected the idea, pressing on to make an electric phonograph that would replace his own machine's hand crank and Volta's foot-treadle mechanism. He continued to view the phonograph primarily as a business machine on which executives could dictate their missives to secretaries. The "Ediphone" and the "Voicewriter" would eventually achieve mass success, but for years they were outsold by the "Dictaphone," also invented down at Volta.

What saved all the talking machines, in the end, were the "phonograph parlors" that opened all over the United States. Customers could speak through a tube and order up a tune. A couple of entrepreneurs named Louis T. Glass and William S. Arnold even invented a prototype of the jukebox, allowing customers to start a phonograph by inserting a nickel. Edison got the message and soon started producing wax musical recordings—though he stubbornly clung to his cylinders over flat discs for years to come. His competition with the companies spawned by Volta would lead to vast and rapid improvements in the talking machine, so that the work of the two giants of nineteenth-century invention would come to dominate home entertainment for much of the twentieth century as well—albeit under such names as the Zonophone, record players, turntables, hi-fi, and stereo, and always as machines that sang and played music more than they talked.

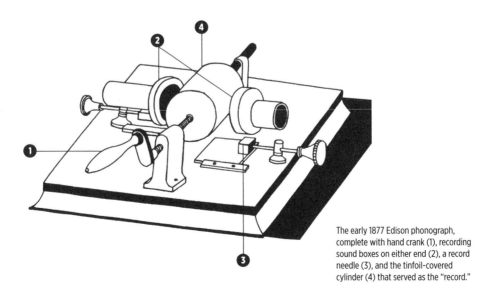

The early 1877 Edison phonograph, complete with hand crank (1), recording sound boxes on either end (2), a record needle (3), and the tinfoil-covered cylinder (4) that served as the "record."

JAZZ

I t is a measure of the insularity of the white, mainstream press in 1920s America that it referred routinely to the portly white composer, violinist, and bandleader Paul Whiteman as "the King of Jazz" and generally implied that he had invented the music.

Whiteman worked with and encouraged black musicians more than most white bandleaders of his time, winning the respect of no less than Duke Ellington, and he deserves kudos from all mankind for pressing George Gershwin to write *Rhapsody in Blue*. But he did not invent jazz.

Who did? Why, all of us, to some degree. The jazz tradition contains traces of music from Irish dances to the Islamic call to prayer as it migrated through West Africa and the Middle Passage. But it comes particularly from African Americans, a people long scorned, oppressed, and persecuted, and not wanted in the New World save in the most silent and subordinate of positions. Their voice was heard anyway, unique and rewarding, what other peoples so frequently think of first when they think of all that is best about us.

"It is America's music—born out of a million American negotiations: between having and not having; between happy and sad, country and city; between black and white and men and women; between the Old Africa and the Old Europe—which could only have happened in an entirely New World," Geoffrey C. Ward and Ken Burns would write. "It is an improvisational art, making itself up as it goes along—just like the country that gave it birth."

Jazz has moved fluidly and freely around America, taking new forms in Storyville and Memphis, Kansas City and Pittsburgh, Harlem and Homestead, Los Angeles and Chicago, the Black Bottom and the Mississippi Delta, and countless towns and cities, riverboats, and fields in between.

Its most prominent point of origin was that strange and wonderful meeting place of at least a dozen American cultures, New Orleans. There, beginning in the first years of the nineteenth century, the rhythms of enslaved black people dancing and playing drums and string instruments in Congo Square began to pour something new into American ears. Their sound began to meld with the music of the city's versatile Creole orchestras and quartets, bringing in Spanish and Haitian airs from Cuba and Santo Domingo; the minstrel shows that took the rudiments of authentic black spirituals and field music—however stereotyped or distorted—around the young country; the funeral bands that played dirges to the cemetery and lighthearted stepping tunes on the way back to chase away the fears of a town racked by death from its surrounding fever swamps.

After the Civil War, the Crescent City's already stunning musical heritage was augmented by many more influences: the brass of marching bands, brought by Irish and Italian immigrants, as well as black Baptist church music, and its cousin, maybe the most powerful American music tradition of all: the blues, that long cry of hope against despair from the Black Belt (see opposite).

These were also joined by "ragging tunes"—improvisations on established white or black tunes that black musicians had been doing already for years. In the 1890s, "ragtime"—"syncopation gone mad," to one critic—swept through the cities of the Midwest where blacks had settled, Chicago and St. Louis, where Scott Joplin's compositions created a national mania for the music that would last for twenty years.

This would be the DNA of jazz, a music that could sound like any number of things, played fast or slow, cool or hot, bebopped or befunked. Infused with all sorts of outside influences, a furious but uplifting group competition, and above all an inspired improvisation, all the same elements that made up the core of the American project in general.

Taking all of it together and making something new—"not spirituals or the blues or ragtime, but everything all at once, each one putting something over on the other," as a contemporary musician put it—was the grandson of a slave who grew up without a father in New Orleans's most integrated neighborhood. Cornetist Charles Joseph "Buddy" Bolden "may have been the *very* first," according to Ward and Burns, to put all these sounds together with his band on New Orleans's Short Street and in dance joints such as Funky Butt Hall.

From there it would be carried all over town—and constantly altered and reinvented as it moved—by the likes of Jelly Roll Morton at the piano of the whorehouses in New Orleans's Storyville red light district. And to the rest of America by a remarkable young man with a cornet, Louis Armstrong, who, after dropping out of third grade and learning to play in a Colored Waif's Home marching band, would become jazz's leading apostle, perhaps known and loved by more people than any other American who has ever lived. And all the way to France by James Reese Europe, leader of the regimental band for the "Harlem Hellfighters," the 369th US Army Regiment that would so distinguish itself in World War I.

The 1920s and '30s would be the zenith of jazz's popularity, perhaps, but on and on the music would go: popularized by Bix Beiderbecke, and Glenn Miller, and Benny Goodman, leader of the first major, integrated band. Turned smooth as silk by Duke Ellington, and Billy Strayhorn's arrangements; set to stomping and jitterbugging by Chick Webb's band up at the Savoy; styled to the songcraft of Ella Fitzgerald and Billie Holiday; turned into bebop by Dizzy Gillespie and Charlie Parker; hard-bopped by John Coltrane, fused by Miles Davis; freed, Latinized, techno-ized, right down to the present day.

To attempt to catalogue even the legends of jazz is to try to name all the grains of sand on the beach—or maybe all the improvisations in a single session. Suffice it to say that it is music that came from all we were and has influenced everything we are.

A CRY IN THE AMERICAN NIGHT
BLUES

He moved like a phantom through our national imagination, glimpsed here and there, slipping across the cotton fields or through the lush Delta underbrush. One moment he'd be there with his guitar, busking for tips in another dusty cotton town. The next moment he would be gone without so much as a word to his friends.

We're still not sure just what day he was born on, or in what year. He lost at least one wife in childbirth, maybe two, and no one's certain what happened to the children. Just who his descendants were was the subject of a years-long court battle, and how he died or where his grave lies is a matter of conjecture. During his brief life he roamed all the way out to San Antonio and up to Canada and New York. He was said to have been so shy that he faced the wall to record his music, though in fact he may have been so musically sophisticated that he was "corner loading," enhancing the sound of his Gibson L-1 acoustic guitar. Many of the great musicians who heard the recordings later were astonished to learn that he was playing alone.

His work was covered by Bob Dylan, members of Led Zeppelin and Fleetwood Mac, Eric Clapton, and the Rolling Stones and Keith Richards, who said of it, "You want to know how good the blues can get? Well, this is it."

He never sold more than five thousand copies of a record while he was alive, but he became a platinum seller almost sixty years after his death and was chosen for the first class of the Rock and Roll Hall of Fame and Museum. It was said that he could imitate any sound, any song on his guitar after hearing it on the radio once, and he could and did play anything from pure country slide guitar to jazz to pop. It was said that he did his practicing in graveyards at night, and who else but Robert Johnson could get scholars to seriously debate whether he had sold his soul to the devil at the crossroads at midnight, and just what the hell that meant? He died near Greenwood, Mississippi, at twenty-seven (probably) after accepting a drink from a poisoned bottle of whiskey, supposedly proffered by a jealous husband after he had already had that cup turned away from him once. When they wrapped up his corpse and took it to a potter's field in 1938, he surely would have been thought the least likely body in the whole of the Mississippi Delta to end up on a US postage stamp, the long, nimble fingers of his left hand poised over a chord change, cigarette dutifully excised from the corner of his mouth.

Only the blues could produce a modern legend as tantalizing as Robert Johnson. If his abilities seemed to develop mysteriously, it was because he was born and raised in a region as rich in America's most resonant folk music as it was wrenching for human existence. Johnson listened and learned there from the likes of such older blues legends as Charlie Patton, Son House, Ike Zimmerman, and countless others.

The blues went back well before them, of course. They went back at least to the call-and-response music of enslaved black people on southern plantations, to their holler songs, spirituals, work songs, and rhyming and repeating ballads; back even to the music and mythology of the Igbo and Yoruba peoples in those parts of western Africa that are Nigeria today.

"Ain't no first blues. The blues always been," maintained the New Orleans clarinetist "Big Eye" Louis Nelson, but they started showing up in southern towns and cities in the years just after the Civil War—freed country blacks, usually penniless, single guitarists much like Robert Johnson decades later, playing for coins on corners. Under more urban and commercial influences, as Geoffrey C. Ward and Ken Burns would write, their music was "stripped to the essentials" and "built on just three chords most often arranged in twelve-bar sequences that somehow allowed for an infinite number of variations and were capable of expressing an infinite number of emotions."

As such, the blues would infuse jazz with its soul. Taken up by W. C. Handy in Memphis, they would be syncopated into ragtime. Taken north to Chicago by Howlin' Wolf, Muddy Waters, and Sonny Boy Williamson, they would become something hard and driving, like the locomotives their rhythms sometimes seemed to imitate. Recorded by the archivists John and Alan Lomax and played by the likes of Huddy "Leadbelly" Ledbetter,

In the tradition of Robert Johnson:
a blues guitarist from the Mississippi Delta.

they would set off the postwar folk revival in Greenwich Village. Taken to New York and George Gershwin, they would become . . . America's very own orchestral music, something altogether sui generis. During their greatest vogue in the 1920s, women blues singers such as Bessie Smith, Ma Rainey, and Mamie Smith became major recording stars, selling hundreds of thousands of records.

Yet there was always some part of the blues that remained unassimilated, defiantly just outside jazz in all its permutations, and irreducibly black. Blues, more than any other American folk tradition, was the music of the scorned, the disinherited, the outsider. It was the music of poor people, played by poor and itinerant people, most often in country roadhouses and juke joints, for little reward or recognition.

Robert Johnson, with his phenomenal talent for mimicry, for innovation, for fitting his original and unique songs into three-minute bites that would fit onto 78-rpm records—then the most popular American musical medium—was a transitional figure, pushing the music he inherited toward what would become the core of rock 'n' roll. It was a tradition that, like so much of black music, didn't come from the devil but from scripture: The last one now shall later be first.

THE GENIUS DETAILS

The origin of the word *blues* is uncertain, but it most likely comes from the seventeenth-century English expression *blue devils*, for depression and/or the hallucinations encountered during withdrawal from extreme alcohol addiction.

Johnson's bass line for "I Believe I'll Dust My Broom" has become a standard rock guitar line.

If the descriptions of Johnson's death are accurate, his death was more likely from some internal ailment or rupture, or possibly bad bootleg whiskey, than from poison.

Johnson's biggest hit in his lifetime, "Terraplane Blues," sold all of five thousand copies.

The Rock and Roll Hall of Fame and Museum named four of Johnson's songs—"Sweet Home Chicago," "Crossroads Blues," "Hellhound on My Trail," and "Love in Vain"—as among the five hundred songs that shaped rock 'n' roll.

THE ELECTRIC GUITAR

L ike all great guitar legends, it starts with a young boy walking down a lonely road, or maybe a railroad track. The boy sees a metal bolt and, wondering, slides it across the strings of the guitar he carries everywhere. Or maybe he drops the guitar right on the steel tracks. Whatever the case, it makes a sound like nothing he's ever heard before.

The boy was Joseph Kekuku, and he wasn't in the Mississippi Delta but on the island of Oahu, around 1885. Soon he was playing his guitar with the back of a pocket-knife, then a polished steel bar, and then replacing his catgut strings with steel ones. One day in 1904, a yacht full of wealthy white haoles from America weighed anchor off the beach where Joseph and his friends were playing. They were so impressed by what they heard that they induced them to sail away with them to Los Angeles and then New York and on to Europe. The Hawaiians took their show, "The Bird of Paradise," all over the world, and Joseph Kekuku never again saw the beaches of Oahu.

If it sounds too good to be true, well, that's show biz. Whatever the particulars, Kekuku really had started a revolution that would transform American music. Believe it or not, the electric guitar, so indelibly associated with rock 'n' roll, came out of a fad for Hawaiian music.

By the 1920s and '30s, Americans were entranced by the music that Kekuku and his bands, and many other Hawaiian artists, such as the great Sol Ho'opi'i, were play-ing in theaters and tent rep shows. Spanish acoustic guitars had been in America for centuries, introduced by the original Hispanic cowboys of the Southwest. (Some of those cowboys were apparently hired and brought to Hawaii in 1832 by King Kamehameha III, which may account for what young Kekuku was doing with a guitar in the first place.)

But the Hawaiian slide steel guitar was something else—a converted six-string acoustic guitar usually played sitting down, with the strings raised above the hollow sound box. Its metallic sound intrigued American country and jazz guitarists, who were always looking for ways to make their instruments heard over all those horns in the big band arrangements of the time. The guitar, it seemed, was just too quiet.

The makers of player pianos and music boxes had already discovered that when metal is passed through a magnetic field the disturbance it causes can be converted into electric current by a coil of wire—the same basic principle behind the telephone (see page 74). In the early 1920s, Lloyd Loar, a Theosophist, vegetarian, and former Chautauqua circuit picker, now working for the Gibson Guitar Company, tried putting a "pickup"—an electromagnet wrapped with up to seven thousand coils of wire—on violas, mandolins, and string basses. The vibrations of the strings were passed through the bridge to the magnet and coil, and then on to an electrical amplifier.

More experiments proceeded apace. In 1928, the Stromberg-Voisinet Company of Chicago came up with an electric guitar on which the vibrations were picked up off the soundboard. John Dopyera, an instrument maker and repairman who had immigrated from what is now Slovakia to Los Angeles with his large family, invented a "resophonic," or "tri-conic" guitar featuring three thin aluminum cone "resonators." A former colleague of Dopyera's, the vaudeville promoter George Beauchamp, working with Paul Barth for Adolph Rickenbacker's Ro-Pat-In Corporation, then developed a method for translating the strings' vibrations *directly* into current—a key breakthrough.

Just who invented the electric guitar depends on how one defines the term, but Beauchamp and Barth were awarded the first patent for one, and in 1932 woodworker Harry Watson turned their design into the "Frying Pan," so named for its long wooden neck and round body.

The Frying Pan was still a lap guitar, though, and nearly all the electric guitars developed until then were still acoustic. Pickups could not distinguish between the vibrations coming from the strings and those from their hollow bodies. They created feedback, jumbled signals, and made it hard to produce sustained notes.

Enter Les Paul, a self-taught acoustic guitarist with a yen for innovation in every phase of the music business. Shortly before World War II he invented several versions of what he called "the Log"—a Gibson Epiphone

The god of the proscenium: the Stratocaster guitar, and the amplifier that makes it heard. The knobs on electric guitars control the tone pickup from the neck and middle, as well as the volume. (Note: some volume knobs will go up to eleven.)

In the 1930s Rickenbacker developed a steel-string Bakelite guitar, Model BD. It would later become a favorite of guitar aficionados, including Jackson Browne's guitarist, David Lindley.

John Lennon and George Harrison played Rickenbacker guitars during their first American appearances and in their early movies. Paul McCartney later used a Rickenbacker bass for much of his career.

The Dobro resonator guitar became a key component in the evolution of bluegrass music. A rare Dopera Bantar—a cross between a five-string banjo and a six-string guitar—was used at times by Bob Dylan in the 1960s.

Leo Fender never took a course in electronics but taught himself the science.

When Leo Fender sold his company to CBS in 1965, he received $13 million for it—or nearly the same amount the network had just paid to purchase the New York Yankees.

acoustic guitar, sawn in half lengthwise and attached to a four-by-four—a big hunk of wood. The Log was a major breakthrough in the development of solid-body guitars— but one whose potential wouldn't be realized for years.

Paul's further experiments with electric guitars nearly cost him his life when he electrocuted himself in his Queens apartment. It would take him two years to recuperate. He put the time to good use and moved to Los Angeles, which was rapidly becoming the center of the guitar industry. There he visited Rickenbacker's shop and became friends with Bing Crosby, who backed his recording experiments. But when Paul took his design for a solid-body guitar to the Gibson company he found little interest.

That would change, thanks to a former iceman, accountant, and bookkeeper named Leo Fender, who had borrowed $600 during the Depression to start a radio repair shop in his native Fullerton, California. During the war years, Fender began working with Clayton Orr "Doc" Kauffman, formerly a Rickenbacker employee, on building their own guitars. By 1945 they were selling a lap steel guitar with a Fender amplifier in a kit. Four years later, Fender produced *his* first solid-body guitar, the "Esquire," to be followed shortly by the groundbreaking "Broadcaster" and "Telecaster" models.

Gibson now scrambled to catch up with its "Gibson Les Paul" model. Further innovations would follow quickly, as the guitar business proved one of the few where all the established companies could both compete and get rich. But the Telecaster was already very much the model of the modern power instrument, and with the advent of rock 'n' roll the music and the instrument were met. The die was cast when Buddy Holly went on *The Ed Sullivan Show* in 1957 and played "That'll Be the Day" and "Peggy Sue" on his Fender Stratocaster. A sound meant to convey the lulling breezes and lapping waves of a tropical paradise would now turn rock guitarists into the strutting, heavy-metal gods of the amphitheater. No one would ever complain again that the guitar wasn't loud enough.

THE AMUSEMENT PARK

I t was not the Statue of Liberty that immigrants headed to New York Harbor first laid eyes on at the turn of the century. Rather, it was a city of fire: Coney Island, where the amusement park first sprang to life.

Seaside boardwalks and amusement piers had begun popping up in places such as Blackpool, England, and later Atlantic City by the mid-nineteenth century. But there was nothing like Coney.

New Yorkers had been traveling out along a shell road to this spit of sand at the edge of Brooklyn since at least 1824 to bathe in the sea, feast on roasted mussels and terrapin at its modest inns, and dance around bonfires on the beach at night. Over the next few decades, more complex amusements and hostelries sprung up, including a 150-foot-tall, thirty-four-room hotel in the shape of an elephant, with shops in its legs and an observatory in its houdah.

George C. Tilyou had a better idea. A native of Coney and a tireless, cunning entrepreneur, Tilyou offered to buy the gigantic Ferris wheel unveiled at the Chicago Exposition of 1893. Turned down, he bought one about half the size, put it up on Coney, and erected a sign advertising it as "the world's largest Ferris wheel."

Noticing the success a primitive aquatic park was having down the street, Tilyou opened his own emporium in 1897—Steeplechase Park. This was the world's first true amusement park of any size, which is to say an enclosed, permanent park with a single proprietor, featuring concession stands, games, and many thrilling rides.

Most of these were as bone-jarring as their names implied: the "Mixer," the "Whirlpool," "Human Roulette," and the "Human Pool Table." The thrill came in the rare opportunity to be thrown together with members of the opposite sex. Tilyou wrapped a half-mile mechanical racetrack around his fourteen acres, where—better yet!—couples had to clutch on to each other to stay aboard their wooden steeds. On exiting, they encountered a surprise: the "Blowhole Theatre," where women's skirts were blown up by jets of air and all were attacked by a deranged-looking dwarf in a harlequin's costume wielding a cattle prod, while their fellow patrons sat and roared with merriment in the "Laughing Gallery."

Here were all the elements of modern, mass entertainment: a hint of sex, the thrill of death, and a lot of wicked fun, laughing at others.

The very first amusement park was technically Sea Lion Park, opened on Coney Island in 1895 by Paul Boyton, an adventurer most famed for traversing the Irish Sea in an inflatable wet suit.

One of the most popular Luna Park exhibits was Dr. Martin Couney's "Infantorium." Couney had pioneered incubators to save premature babies but could not get a major hospital to use them. So he set up shop for forty years on Coney Island, charging customers twenty-five cents admission. Couney saved 7,500 of the 8,500 premature babies brought to him and saw incubators finally adopted by hospitals shortly before his death in 1950.

Benches on Coney were routinely electrified to give customers a mild *zetz* and get them up again and spending money if they tarried too long.

Scientific American estimated that by 1910 Coney Island generated enough power to light a typical American city of five hundred thousand.

Coney Island's biggest crowds arrived with the subway, starting in 1921. The largest estimated crowd at Coney Island ever was two million people—one-quarter of the population of New York—on the Fourth of July in 1947, a scene famously photographed by Weegee.

Steeplechase Park burned down after a few summers, but Tilyou rebuilt it and enclosed it in a painted glass trellis. (His rivals said he went to church on Sunday to pray for rain.) He was competing with the singular genius of Frederic Thompson, a whimsical showman and architect, who was convinced that "buildings can laugh quite as loudly as human beings." Thompson set out to do just this in creating Luna Park, easily the most beautiful amusement park ever constructed—a Seussian dreamscape of fantastical towers, minarets, obelisks, and spires not quite like buildings found anywhere else on earth.

"A spirit of frolic must be manufactured," Thompson maintained, in the ethos of the machine age, "and it cannot dwell where straight lines, dignified columns, and conventional forms dominate."

Into this fantasia he shoved twenty-two acres of lagoons, aerial rides, spinning half-moons, elephant and camel rides, endless garlands of flowers, leering pig and wolf and clown heads, all sorts of roller coasters and "dark rides," such as "A Trip to the Moon" and a visit to Venice, and above all—electricity.

Electricity was what ended up mesmerizing even the most cynical visitors to Coney. In this age before neon, Luna Park alone boasted 1.2 million lightbulbs, outlining nearly every dazzling inch of it once the sun set. Dreamland, a third great park added in 1903, tossed another one million lightbulbs up on the night skyline, and its powerhouse became its own exhibition, featuring murals on electricity and beautiful gold and enamel instruments.

In and around these three marvelous parks was a constant cacophony of bands, animal shows, gigantic dance halls, vaudeville theaters, and more tawdry entertainments along Coney's very own "Bowery"; vast eateries, such as Feltman's "breezeway," where the hot dog was supposedly invented; and the world's very first roller coasters, "the aristocrat of park attractions," as historian Richard Snow called them.

There were, as well, any number of "educational" rides and exhibits, re-creations of recent events, and grand tableaux, including displays of exotic tribal peoples in their "native" habitats. Visitors could see what it was like (more or less) in the bowels of a coal mine, or at the bottom of the sea, or high in the Swiss Alps. They could see premature infants literally struggling for life—or actors diving out of a burning city tenement, much like the one they had often just come from. They could, in short, see the drama of their own lives, and their own world, played out splendidly before them.

Dreamland would burn down in 1911, ending the golden age of Coney Island. But by then "Coney Island" was already synonymous with great, gaudy fun, and hundreds, maybe thousands, of "electric parks" or "trolley parks" were built around the world—amusement

The entrance to Luna Park, "the heart of Coney Island" and the most beautiful amusement park ever built, circa 1903.

parks deliberately placed at the end of electric trolley lines to promote their use, and land development along the way.

Steeplechase, the last of the three great original parks, closed in 1964. The developer Fred Christ Trump, father of Donald, held a big party. At the stroke of midnight he gave out bricks to all the guests and invited them to hurl them through George C. Tilyou's beloved glass trellis, with its iconic, grinning heads and the famous motto, "STEEPLECHASE—FUNNY PLACE."

Coney Island would live on in a humbled state, but its glorious past can still be glimpsed, maybe here and there on those summer nights when the moon is half full. In the words of Maxim Gorky: "Thousands of ruddy sparks glimmer in the darkness, limning in fine, sensitive outline on the black background of the sky, shapely towers of miraculous castles, palaces, and temples. . . . Fabulous and beyond conceiving, ineffably beautiful, is this fiery scintillation."

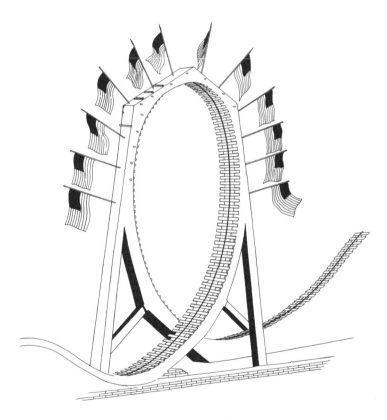

The roller coaster, crowning glory of the amusement park, was invented at Coney Island. The Flip-Flap Railway, an early model depicted here, was a perfect circle instead of an oval, and as a result subjected most of its patrons to whiplash.

THE POLO GROUNDS

T
he play doesn't seem that hard if you don't know what you're looking at. A streak of a man, racing toward a wall. The ball comes down, and he catches it, almost routinely, over his shoulder, whirling and throwing it back toward the camera in one motion. It's only when you know where he was that you understand it is one of the greatest plays ever made by an athlete.

And one that could only be made at the Polo Grounds.

America's original, professional baseball parks were hastily constructed wooden grandstands, with just enough amenities—and enough enclosure—to let teams charge people money to get inside. Gradually, their designs became more sophisticated until, between 1909 and 1923, at an average cost of $2 million, nearly every major-league team built or moved into new steel-and-concrete structures (all paid for entirely by the club owners with no public funds).

There were bigger arenas. By the 1920s, college football stadiums could already seat over seventy thousand fans, while most ballparks still had capacities of less than fifty thousand. But the college stadiums were just big, interchangeable ovals with the same precisely measured gridiron in the middle.

Baseball parks were still idiosyncratic, shaped by the urban environments around them. Products of the machine age, they were nonetheless the last stand of the particular, the unique, in American commercial life—wedged into street lots, built around factories, stables, and houses; set next to streetcar lines or railroad yards that billowed smoke over the outfield walls.

Each baseball field was different and helped shape the nature of the team it housed. Brooklyn's Dodgers got their name because of the trolleys that players and spectators alike had to dodge to get to their old Washington Park field. Washington's Griffith Stadium had a sudden right angle in dead centerfield—a bulge where five homeowners had refused to sell. There was the ivy that still clings to the brick walls of Chicago's lovely Wrigley Field; the great, green wall looming over left field in Boston's Fenway Park; the white filigree hung like a wedding cake's decoration from Yankee Stadium's upper deck.

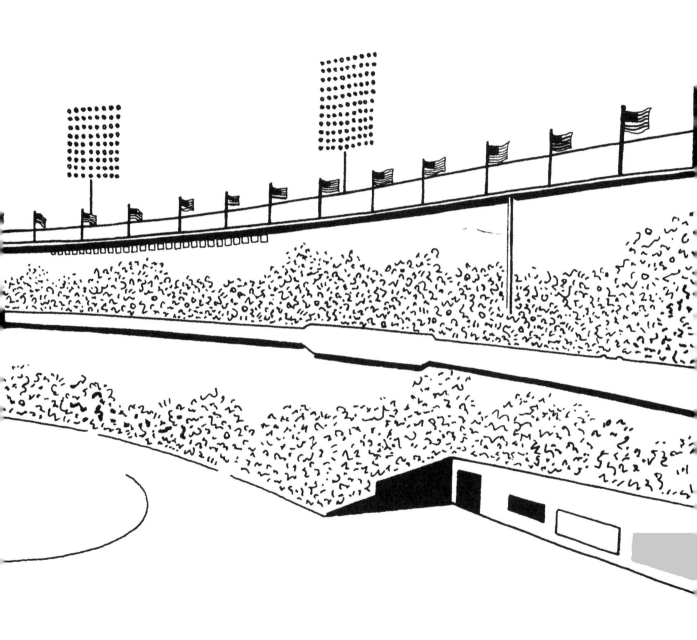

The Polo Grounds as seen from the right-field bleachers,
near where Mays made his famous 1954 catch.

Before the Polo Grounds was fully enclosed, Broadway actors and actresses often traveled there by carriage or motor car to park their vehicles and watch the afternoon games from centerfield.

The song "Take Me Out to the Ball Game," by Jack Norworth and Albert Von Tilzer, was inspired by an ad for a game at the Polo Grounds that Norworth saw in the subway (see page 21), during the tumultuous pennant race of 1908.

The Giants abandoned the Polo Grounds after the 1957 season when they moved to San Francisco. It would be used by the New York Mets during their first two seasons of existence, 1962–63, and then demolished in 1964.

Among its many other quirks, the Polo Grounds by 1947 also had an apartment onsite—for groundskeeper Matty Schwab, under the left field stands. His son, Jerry, used to hold sleepovers with his friends in the outfield.

There was the little hillock in left field in Cincinnati's Crosley Field, which once sent Babe Ruth tumbling; the turreted Victorian tower at the entrance to Philadelphia's Shibe Park; the "Jury Box" in Boston's Braves Field; the Coop and the Crow's Nest at Pittsburgh's Forbes Field. The ticket booths looked like boathouses, or beach pavilions with fringed awnings. Huge, hand-operated scoreboards showed the progression of every other game around the country, inning by inning. It was these parks that instilled baseball forever in the American psyche, left generations remembering the thrill they had when they first walked out in the stands and saw that stunning sweep of green before them—the cliché that never grew old.

The most eccentric ballpark of all was the Polo Grounds, the home of the New York Giants in upper Harlem. It was a strange place where strange things happened. The only man ever killed on a major-league playing field, Ray Chapman, died at home plate, hit in the head by a "submarine" pitch. A fan once died in the stands, struck by a stray bullet fired from a neighborhood rooftop, killed instantly as he turned to say a word to his companion. The game went on, and other fans fought for his seat.

No one ever played polo there; the name was carried over from the Giants' previous digs at 110th Street and Fifth Avenue, demolished in 1889 when the city built 111th Street through what had been the outfield. The new park the team built for itself at 155th Street and Eighth Avenue was shaped like a gigantic horseshoe, facing the Harlem River, on land recently dredged up from the water. Above it towered Coogan's Bluff, or "Dead-Head Hill," where those fans who couldn't get tickets crowded to watch what they could of the game. They would become a fixture in newspaper photos: forlorn-looking men in dark suits and bowlers, a flock of envious crows peering down at the festivities.

The Osborn Engineering Company of Cleveland, which built several of the most memorable baseball parks, decorated the Polo Grounds with the usual elegant touches of the time. Along the facade of the upper deck was a frieze of ballplayers in action, and shields representing all eight National League teams hung from the roof. There were box seats modeled after the emperor's box in the Roman Colosseum, and iron scrollwork in the shape of the Giants' "NY" emblem on each aisle seat.

The Polo Grounds featured the most extreme dimensions of any park ever built. It was almost not a real ballpark at all, but a baseball diamond transposed on a dreamscape. Down the lines were the easiest home run shots in the big leagues, 279 feet to left and 257 to right. The outfield walls then undulated out to a centerfield 483 feet from home plate. There, filling the gap in the horseshoe, was a building several stories high, housing the team offices and clubhouses for both teams. A little balcony and two sets of stairs led to the field from the clubhouses. It was here that Bobby Thomson would acknowledge the rapturous applause of the fans after hitting the most famous home run in baseball history, his 1951 "Shot Heard 'round the World." Here was where the Giants would run for their lives to the clubhouse just six years later, following their last home game in New York, chased by their bitter fans after it was announced that they were leaving for San Francisco.

The Polo Grounds' dimensions would lead directly to two of the most dramatic moments in baseball history. One of them was Thomson's home run into the left field stands. The other was the catch made in the 1954 World Series by Willie Mays, sprinting back across an outfield so deep it seemed to outstretch any human capacity. Off with the crack of the bat, pounding his glove once, twice, as he raced toward the wall nearly five hundred feet from home plate, Mays pulled the ball in over his left shoulder with barely a shrug, even as the fans rose to their feet in the bleachers, mouthing the words, "Oh, my God!" He stopped himself, somehow, turned and windmilled that throw back toward us—imprinting himself indelibly in our memories, in the only ballpark that ever possessed the capacity to show us the full measure of his beauty.

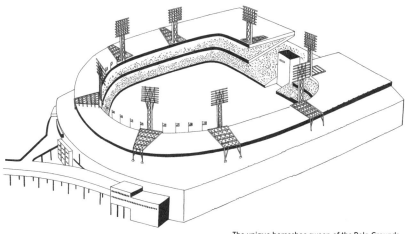

The unique horseshoe sweep of the Polo Grounds, with the clubhouse and team offices in center field.

WONDERS AND ATMOSPHERICS
THE MOVIE PALACES

"**W**e sell tickets to theaters, not movies," Marcus Loew, a pioneer of American film, liked to say, and it was true that the American studio system was in many ways an inverted real estate empire, its magnates moving into movies to ensure content for their properties. Many of the early studio heads, including the likes of Loew, Louis B. Mayer, Adolph Zukor, and the Balaban brothers, were building and running movie houses before they ever made a film.

Yet there was always more to it than that, a romance with America that even the flintiest of the movie moguls—a very flinty bunch—felt in their souls. Like the movies they showed, the great theaters began as the dream fulfillment of immigrants from all over Europe, and in their cinematic empires it was nowhere better realized. Even the best studio turned out clunkers and flops, and film was evanescent. The theaters were palpable and never less than spectacular.

The first real movie palace was the Mark Strand Theatre, opened in 1914 at Broadway and Forty-Seventh Street in Manhattan and designed by the Scottish immigrant Thomas Lamb, who had been educated as an architect at Cooper Union's free university for working people. The Strand seated nearly three thousand patrons and was a gaudy, baroque marvel.

The idea of the palace really took off in Chicago, where A. J. Balaban, his brother Barney, and partner Sam Katz began to build movie theaters on a scale never witnessed before. Sons of a poor Jewish immigrant grocer and his wife, A. J. and Barney convinced their parents to help finance their acquisition of a floundering Chicago nickelodeon called "the Kedzie" in 1908, for $178. The Balaban brothers got into motion pictures reportedly because their mother, Gussie, came home from her first movie and told them, "The customers pay before they even see what they're paying for! There'll be money in that business!"

Soon they were building the "big three" of Chicago movie palaces—the Uptown, the Chicago, and the Tivoli. Each would include the standard features of the movie palace: auditoriums and balconies that held three thousand to five thousand patrons, and onstage orchestras of fifty to sixty musicians. They also featured top *live* acts between screenings, such as the Marx Brothers, Sophie Tucker, Ginger Rogers, and

The former Loew's Paradise Theater, preserved today on the Grand Concourse of the Bronx.

At the movie palaces, you were attended by an army of immaculately groomed and uniformed ushers. At the Balabans' Tivoli they even put on military-style drills, taught to them by Marine sergeants.

Manhattan's Roxy Theatre, "the Cathedral of the Motion Picture" and the largest of all the old movie palaces with nearly six thousand seats, was named for impresario Samuel "Roxy" Rothafel and featured high-kicking chorines called "the Roxyettes," who would soon move on to Radio City Music Hall and become "the Rockettes."

Balaban installed the first system in an American movie palace by having a large fan blow air over blocks of ice, in the basement of the Central Park Theatre, in 1917. The system was very loud and sometimes blew water up over the customers, but it was still a success during the silent era.

A. J. Balaban's innovations included free admission for children under ten and a curbside "baby carriage service" where mothers could check their infants. They would be alerted by slides on the side of the screen informing them, "Mother Number 56, your baby is crying."

the bands of Paul Whiteman and John Philip Sousa, as the movies continued to evolve from being just one more vaudeville act.

The Balabans built dozens of only slightly smaller theaters in the outer neighborhoods of Chicago. This was the reversal of the revolution of electricity and architecture (see page 135) that had brought the masses to the urban center. The movies were now bringing validity to the 'burbs, becoming nodes of sophistication around which they would grow.

By the 1920s, Americans routinely spent five or six hours on a Saturday or a Sunday "at the movies," watching two features as well as a newsreel, coming attractions, shorts, live acts, and a cartoon. But the movie palaces themselves were as good as any show. You could sit amid the splendor of any number of exotic places and cultures: Mediterranean or Spanish Gothic, Babylonian or Aztec, Italian Renaissance or Egyptian (particularly after the discovery of King Tut's tomb in 1922), often jumbled together as wildly and thrillingly as the dreamlike amusement architecture on Coney Island (see page 229). "The theatre should be a veritable fairyland of novelty, comfort, beauty and convenience," said A. J. Balaban—insisting again on that seemingly impossible American contradiction of "manufactured fun" that was somehow both thrilling and reassuring.

By the end of the 1920s, each of the "big five" movie studios, Loew's (always pronounced "Low-eez," at least in New York), Paramount, Warner Brothers, RKO, and Fox, owned their own chains of palaces around the country. They were, in many ways, the ultimate sets of the studio system, erected in as little as two months, their furnishings often concoctions of plaster and paint, ingeniously crafted to seem like marble and carved wood.

Breaking out of the Chicago school of opera-house movie palaces was another immigrant, John Adolf Emil Eberson, a character nearly as fabulous as his buildings. Eberson earned a degree in electrical engineering at the University of Vienna, fought several duels, joined a Hussars (cavalry) regiment, and was jailed after assaulting

The lavish ornamentation of the Loew's Kings Theatre in Brooklyn, now restored to its former glory.

an officer. Escaping from prison, he made his way to the United States, where he became, according to theater historian David Naylor, "an architectural Johnny Appleseed for Sunbelt theater-goers."

Beginning in 1920 with the Hoblitzelle Majestic Theatre in Houston, Eberson started building "atmospheric theaters" designed to make audiences feel as if they were seated in an exotic outdoor setting—an idea that usually started not with a drawing but with a story. "We visualize a dream, a magnificent amphitheater, an Italian garden, a Persian Court, a Spanish patio, or a mystic Egyptian temple yard, all canopied by a soft, moon-lit sky," he wrote.

"He had a flair for the dramatic," remembered his daughter, Elsa, who, along with Eberson's wife and other children, worked devotedly beside him to create all the magnificent frippery of his palaces.

To get a sense of what it meant to enter an Eberson creation, one need only consider the Bronx's beloved Paradise, one of five Loew's "Wonder Theatres" in the New York area, so named for their thunderous Robert Morton Wonder Organs. Finished in 1929 at a cost of $4 million, the Paradise featured a five-story facade of cream-colored terra-cotta and red Levanto marble, with a crowning equestrian statue of St. George slaying a dragon who breathed fire every hour on the hour. Inside, one walked past classical statues, exotic birds in real shrubs (in fact, they were stuffed pigeons), enormous chandeliers, a grand piano (and pianist), a fountain with live goldfish, walls upholstered in salmon-colored silk, copies of famous oil paintings, urns, cherubs, caryatids, parchment lanterns decorated with Chinese calligraphy, phone booths "carved to resemble sixteenth-century sedan chairs," and allegorical murals of Sound, Story, and Film.

And that was just getting to the auditorium: a near-perfect replica of a Mediterranean courtyard on a beautiful evening. A Mediterranean courtyard, that is, lined not only with artificial shrubs, vines, and cypress trees but also with statues or busts of Shakespeare, King Arthur, Wagner, Venus, Apollo, and Cupid. Above was a romantic, azure-blue sky that would slowly darken as showtime approached, its stars twinkling—thanks to the temperamental "Brenograph" machine, which the Ebersons also sold to theaters, along with its "universal electric motor with variable speed control with fleecy cloud effect complete in case," for $290.

The Paradise sat smack in the middle of the Grand Concourse, the main boulevard of the Bronx, modeled after the Champs-Élysées, and the promised land for all the penniless immigrants and their children who had made it into the American middle class. By the year it was built, the golden age of the movie palaces was nearly over, but the dream—and the reality—they had provided would live on.

Indeed, many of the old movie palaces survive, often the one building in your town or city to make the National Registry of Historic Places. They are usually considered too big—or too splendid—to show films today, instead featuring rock shows, concerts, plays, and other live performances, fleecy effect not included.

TELEVISION

Nothing else engaged inventors' minds over as wide a swath of the world as the idea of television. Depending on how you count it, the first image was televised in 1909, or 1911, or 1925, or 1927, by inventors in France, or Russia, or England, or Japan, or the United States. In 1931, CBS's first TV station, W2XAB in New York City, began nearly two years of regularly scheduled broadcasts with a show featuring George Gershwin, Kate Smith, and Mayor Jimmy Walker.

The trouble was, this was all "mechanical" or at best "electromechanical" television. Light would carry images through the perforations in a spinning "Nipkow disk" and strike selenium photocells, which converted it into electrical impulses. These impulses would then be carried by wire to a second spinning disk, which would turn them *back* into images and project them onto a screen. It was an ingenious process, but it produced blurry, curved, flickering, and nearly indistinguishable images, sometimes no bigger than the size of a postage stamp.

A better way was found by an enterprising young American, born in a Utah log cabin. Philo Farnsworth was six years old when he first saw a telephone and a gramophone and decided that he would invent things. When he was twelve, his family moved to his uncle's ranch in the Snake River Valley, and whole worlds opened up to

An early electronic television from the 1930s.

him. There was a cache of old science and technology magazines in the attic, and a Delco battery that gave the ranch electricity, and then there was the land itself.

Farnsworth used the battery to electrify his mother's hand-powered washing machine and won a twenty-five-dollar prize from a pulp magazine for inventing a magnetized car lock. He got up at two every morning to work or read in his attic laboratory before doing his farm chores and riding off to school on a horse. One day, he stared out at a newly harrowed field and saw in the marks he'd made across the land the path of his future.

The furrows reminded him of something he'd read by a Scottish inventor, A. A. Campbell-Swinton, in an old copy of *Nature* magazine. When he was just fourteen, Farnsworth drew it out on the blackboard for his high school science teacher, Justin Tolman: plans for a wholly electronic television. Images would be captured and conveyed by electron beams between cathode ray tubes, darting back and forth like all those lovely furrows Farnsworth had glimpsed in an Idaho beet field. Campbell-Swinton had dismissed it as all but impossible to realize. Farnsworth didn't think so. Nor would his amiable California backers, Leslie Gorrell and George Everson.

"If we were able to see people in other countries and learn about our differences, why would there be any misunderstandings? War would be a thing of the past," Farnsworth would say, laboring under the common illusion that the more people got to know each other, the better they would get along.

It was poor philosophy but good woo. Farnsworth married his high school sweetheart, Elma "Pem" Gardner, when he was twenty, she just eighteen, and struck out for California. Pem's brother and Philo's best friend, Cliff, joined them in their San Francisco boardinghouse. Pem made drawings of Philo's work. Cliff took a course in glassblowing and created what would become the universal shape of the TV receiving tube, with its flat end for a screen. Philo's sister Agnes supported everybody with a job at the phone company. Gorrell and Everson did their part, winding countless yards of wire for magnetic coils and trying to dig up more investors. Gorrell used to come in to Farnsworth's small lab at the end of a day and playfully slap him on the back: "Hi, Phil, haven't you got that damn thing to work yet?"

On August 30, 1927, Farnsworth thought he saw a fuzzy image of a horizontal line on the screen. To be sure, he took the whole apparatus apart, rebuilt it, and tried it once more, eight days later. There was the line again—the glass slide of a triangle, "broadcast" from the next room. Farnsworth rotated the slide. The image rotated. With Everson, he sent a wire to Gorrell, away in Los Angeles: "The Damned Thing Works." He was twenty-one years old.

Farnsworth's "image dissector" captured images, broke them down, then re-created them through an "image oscillite," another cathode ray tube. The spinning Nipkow disks were replaced by caesium, which emits electrons when exposed to light. People looked like people instead of flickering ghosts.

Over the next few years, Farnsworth perfected his television, doing away with its motor generator so that it became wholly electronic, with no mechanical parts. To raise money for his invention, he and his backers went to the press and put on public demonstrations. One of the individuals they interested was David Sarnoff, who had gone from a penniless, immigrant newsboy to the driving force behind RCA, RKO, and the new NBC radio networks before he was forty.

When Farnsworth refused to sell out to him, Sarnoff did his best to destroy him. He pressured companies such as Philco to end their agreements with Farnsworth and his backers and sent one of his employees, Vladimir Zworykin—who had tried for years to invent electronic television and failed—to see Farnsworth. When Farnsworth and Cliff Gardner built an image dissector in front of him, Zworykin remarked, "This is a beautiful instrument. I wish I had invented it myself."

Sarnoff went ahead and tried to say he had anyway— not the last time he would try to steal an idea from a great inventor. But Farnsworth's lawyers had a surprise waiting. They had found his old high school teacher, Justin Tolman, who produced a ragged piece of notebook paper recovered from his attic. It was a copy of the electronic television Farnsworth had sketched out for him when he was fourteen years old.

Sarnoff finally paid up—$1 million plus a royalty on every set sold for the duration of the fight. But the extended struggle changed something in Farnsworth, as had the death of his eighteen-month-old son. Farnsworth was quickly disillusioned by what he saw on the tube. His wife remembered that for a time he would not even let the word *television* be spoken at home.

His family helped bring him back from despair. Watching the moon landing (see page 37), he told Pem, "This has made it all worthwhile." He returned to his latest obsession: trying to develop nuclear fusion as an energy source.

"He thought fusion could be used to eliminate pollution and help save the planet," Pem remembered. "That's what he was working on up until the day he died in 1971."

(see page 37)

THE GENIUS DETAILS

Mechanical televisions initially produced only 18 horizontal lines, and no more than 240. Farnsworth's produced 500 lines. Standard televisions today produce 525 to 625 lines; high-definition TVs, 1,125 to 1,259 lines.

Movie stars Douglas Fairbanks and Mary Pickford were among those to visit Farnsworth's San Francisco lab. A loose wire accidentally distorted the images of Pickford and Fairbanks and left them convinced that television had little future.

By 1946, only 0.5 percent of all American households had a television set, with some three-quarters of these located in the New York metropolitan area. By 1954, 55.7 percent of US households had a set; by 1962, 90 percent did.

The first US television station, W3XK, was licensed in Wheaton, Maryland, in 1928.

David Sarnoff broadcast President Franklin Roosevelt's opening remarks at the start of the 1939 New York World's Fair, presenting television as an RCA creation.

THE VIDEO GAME

T oday, over half the population of the United States spends at least one hour a day playing video games. But it wasn't always that way. There was once a time when people spent most of their waking hours staring at screens to watch dramas, news programs, documentaries, and other old-fashioned entertainment quaintly known as "content."

They could have been playing video games, too, as the first video game was actually invented in 1947, coinciding perfectly with the advent of scheduled commercial television programming. All right, it wasn't *really* a video game but a "cathode ray tube amusement device," according to the irresistibly sexy name that its inventors,

South Carolina native Thomas T. Goldsmith Jr. and the mysterious Estle Ray Mann, put on their patent application. But it *was* undeniably a game you could play on a screen, inspired by the work that Goldsmith had done working on radar console screens during World War II.

The game worked like an Etch A Sketch. Beams from cathode ray tubes appeared on the TV screen as dots (just as, with radar, they had represented the reflected radio waves that indicated incoming planes). Players had a few seconds to make the dots reach an airplane (painted on a screen overlay) with the use of some buttons, and then "shoot it down." No actual video signals were being generated, but Goldsmith even managed to simulate an explosion by having the cathode ray defocus when it hit the plane successfully.

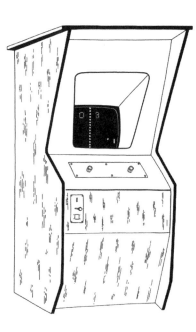

Pong: invented by scientist Willy Higinbotham as "Tennis for Two" in 1958, developed into the home video game "Table Tennis" by Ralph Baer in 1966, and made into the queen of a thousand arcades by Atari's Nolan Bushnell in 1972.

Goldsmith, though, was too busy developing America's first television network to exploit his invention. A few years later, a Cambridge doctoral student in computer science named Sandy Douglas invented the world's first "computer game," "OXO," a game of "Noughts and Crosses" (i.e., tic-tac-toe). "OXO" was generated by EDSAC, the world's first storage-program computer; players used a mechanical telephone dialer, and the results showed up on a pixeled screen. But the screen was still generated by cathode ray tubes, and the computer, which filled up a large room, ran on vacuum tubes. It was like using a couple of redwoods to play a game of tennis.

Speaking of tennis, Willy Higinbotham, a physicist at Brookhaven National Laboratory, was looking over instructions on how to plot trajectories and bouncing shapes on a Donner Model 30 analog computer when he realized, "Hell, this would make a good game." In the space of about four hours, Higinbotham designed just that, "Tennis for Two." An oscilloscope generated a simulated side view of a tennis court, with players able to press a button to hit the "ball" and twist a knob to direct its angle.

"Tennis for Two" was an immediate sensation—on the lab's visitors' day. It made a sound when you hit the ball and could even be configured to simulate playing tennis on Jupiter or the moon.

"The game seemed to me sort of an obvious thing," Higinbotham later remembered. He was doing government work at the time, so "if I had realized just how significant it was, I would have taken out a patent and the U.S. government would own it!"

Good-bye, federal deficit. Instead, "Tennis for Two" was soon cannibalized for parts in more serious work.

Seriousness would prove as usual to be a nearly insurmountable barrier against progress. Ralph Baer, born Rudolf Heinrich Baer, was an eleven-year-old Jewish boy forced to leave his German public school after Hitler came to power. His family immigrated to the Bronx, where at sixteen Baer went to work ten hours a day, sewing leather cases for manicure kits in a factory. He learned to repair radios through a correspondence course, then served as an intelligence

In 1961, MIT computer scientists Steve Russell, Martin Graetz, and Wayne Wiitanen invented the "Spacewar!" computer game, but it required computer hardware costing hundreds of thousands of dollars.

Atari's "Pong" game was developed by engineer Allan Alcorn, who had been given the task as a training exercise by Bushnell.

The original Odyssey package consisted of a master control unit, two player control units, color overlays for the television screen, and electronic cards for a dozen different games—plus a pair of dice and a pack of playing cards. Baer soon developed a light gun for four more games that were also sold in the Odyssey package. Another six games were sold separately. The first game Baer invented for Odyssey consisted of one square chasing another. Quiz, shooting, ball-and-paddle, and other chase games were also included.

Five million Americans spend at least forty hours a week playing video games—the fastest-growing modern media creation, going from $10 billion in worldwide sales in 1990 to over $80 billion today. Gamers spend roughly $8 billion a year in real money, buying virtual items for their games.

officer in Europe during the war. He spent his spare time turning German mine detectors into radios so soldiers could listen to music.

After the war, Baer used the GI Bill to get one of the first degrees ever awarded in television engineering. Designing a TV set for Loral Electronics in the Bronx in 1951, he suggested adding a game-playing feature and was told, "Forget it. Just build the damn TV set; you're behind schedule as it is."

A modern gaming console, able to create whole worlds of entertainment on one's own screen.

Fifteen years later, waiting for a colleague on a dripping summer day outside New York's Port Authority Bus Terminal—circumstances ripe for generating desperate thoughts—Baer had an epiphany and immediately began sketching plans for a "game box" that would work with any American TV set. This time his employer, military contractor Sanders Associates in Nashua, New Hampshire, was more open-minded and ponied up $2,500 for Baer to build a console with engineers Bill Harrison and Bill Rusch. After working through six prototypes, they developed "the Brown Box," so called for the wood-grain vinyl adhesive applied to make it look more appealing to investors. A key addition was Rusch's suggestion that a third "dot" be added so that the Brown Box could generate a game they called "Ping-Pong."

When he set up his prototype on a television in the US Patent Office, Baer remembered, "Within 15 minutes, every examiner on the floor of that building was in that office wanting to play that game." Compared to modern video games, Baer's was fairly primitive. It had no software, for one thing—forty transistors and forty diodes ran everything—but it was the very first home video game, sending video signals to a television set.

When it was marketed finally as Magnavox Odyssey in 1972, its sales were disappointing—just 330,000 units in four years, thanks in part to Magnavox charging $100 an Odyssey (Baer suggested $19.95) and limiting distribution to its own dealers.

Entrepreneur Nolan Bushnell, the founder of Atari, would soon surpass Odyssey, marketing a wildly popular arcade video game known as "Pong," and Bushnell would bitterly contest Baer's title as "the father of video games" for decades to come. But the courts ruled that "Pong" was indeed derived from the Ping-Pong game Bill Rusch had suggested, and Magnavox cleaned up against all other patent infringements as video games swept across the world.

From the mind of the eleven-year-old boy the Nazis didn't want in school would flow Xbox, PlayStation, Wii, and doubtless many games yet to come. The television industry is just grateful that Baer waited a few years while they were building their sets.

FOOLING THE MIND
THE VIRTUAL ENVIRONMENT

I t could be said that humans have been chasing after "virtual reality" ever since the first stereoscopes appeared in the mid-nineteenth century. A pair of prismatic lenses focused on a postcard held at the length of a stick, stereoscopes and stereopticons created a mild 3-D effect, attempting to put the viewer "in" a city or landscape.

The movies turned to 3-D in the 1950s and '60s, trying to give customers something they couldn't get on their TV at home, and an array of low-budget schlockmeisters figured, why stop at one sense? Moviegoers got to experience the likes of "Smell-O-Vision" and "Aromarama"; "Illusion-O" (in which viewers could choose to see or not to see onscreen ghosts); "Percepto" (in which an electric buzzer was placed under seats); "Emergo" (featuring a twelve-foot plastic skeleton that emerged from a box by the screen and swung out over the audience on a wire); and "Bloody Vision," in which employees in monster suits ran out into the audience and "abducted" young girls.

A more serious effort was put forward by New York native and World War II veteran Morton Heilig, a respected cinematographer, who became known as "the Father of Virtual Reality." Intrigued by "Cinerama," in which three cameras projected a movie onto a wide screen, Heilig noted that the process still engaged only 72 percent of the viewer's vision. Extolling "the Cinema of the Future" and "Experience Theatre," Heilig invented the "Sensorama Stimulator" in 1962, a sort of arcade booth in which patrons saw five short films on a wide screen, shot in different tracks so that they could simultaneously hear stereophonic sound, rock and vibrate, feel the wind in their hair, and smell real odors.

There were no takers for Heilig's prototype, though, and advances in virtual reality shifted to attempts by the Air Force and NASA to simulate flight experiences. Toward that end, Ivan Sutherland, a Harvard professor who had already invented the sketchpad, and his student Bob Sproull created the first virtual reality head-mounted display (HMD) helmet in 1968. It was a huge step forward but an awkward one. The device Dr. Sutherland and Sproull came up with was too heavy for anyone to lift and had to be suspended from the ceiling, giving it its name, "The Sword of Damocles"—hardly a reassuring moniker. It featured head tracking—following the user's gaze, a vital part of virtual reality—but only wire frame models (i.e., outlines of landscapes).

In the years that followed, virtual reality would proceed in fits and starts to "goggles and gloves" systems, multiplayer arcade games, and cubic immersive rooms where players could see their own bodies and those of people around them. Many of these advances moved from the arcade to home computers and televisions and proved increasingly intricate. But they still seemed more like highly advanced simulations and gaming spaces than another reality.

Again and again, computer experts predicted an imminent breakthrough. "Affordable VR by 1994," claimed *Computer Gaming World* in 1992. In 2014, Michael Abrash, the newly hired chief scientist at Oculus VR, predicted that a great virtual reality system would be available by the following year or very soon after.

Sadly, no. And the whole idea that a completely immersive world—something like the holodeck in *Star Trek*—was just around the corner became so irksome that virtual reality innovators started trying to dial down expectations, renaming what they were trying to make a "virtual environment."

The problem lies more in ourselves than in our technology. Human beings are incredibly sophisticated creatures who evolved in the treacherous environment of the African savanna, needing to instantly distinguish predator from prey. We're very hard to fool. A latency period—lag time—of more than just fifty milliseconds between when we move our eyes to look at something and when it appears signals to the brain that it's a simulation and can cause "swimming" or physical nausea much like seasickness. Adding all the nuances we routinely pick up can be incredibly time consuming and expensive. Right now, it can take a team of programmers over a year to accurately re-create one room in virtual space.

Entering even the most advanced virtual environment today requires, among other things, donning helmets or visors, gloves, and customized biosensor suits—surefire reminders that we are not in a natural state. The ability to imprint images on our eyes and/or brain chips may change this some day—but that would be in a world most humans would be very wary of entering.

"The ultimate display," according to Dr. Sutherland, would "be a room within which the computer can control the existence of matter"—a comment that conjures up the sentient, and murderous, computer HAL

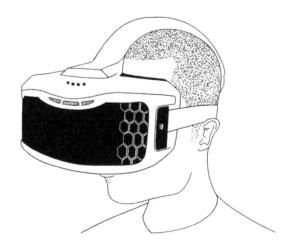

The latest headgear designed to create a "virtual environment."

in Kubrick's *2001: A Space Odyssey*. Yet the next generation of virtual environment systems, with names like something out of the science fiction where virtual reality has long been much more successful than it is here—Oculus Rift, Sulon Cortex, Avegant Glyph, Altergaze, Durovis Dive, Project Morpheus—continue to thrive and to be snapped up for small fortunes.

Many dismiss the preoccupation with entertainment in developing a virtual environment, pointing to the uses to which the technology is already being put in surgery, physical and mental therapy, manufacturing, urban design, historical simulations, and military training. But one should never bet against fun. Just ask the inventors of "Bloody Vision."

The French playwright, poet, actor, and director Antonin Artaud first coined the term *virtual reality* back in 1938—though he meant it more to refer to the willfully suspended reality of live theater.

The term *virtual reality* was popularized in 1987 by Jaron Zepel Lanier, a pioneer in the field.

Oculus Rift inventor Palmer Freeman Luckey began to experiment with high-voltage electronics projects when he was just eleven years old and Kickstarted his company, Oculus VR, raising $2.4 million, in 2012, when he was just twenty. Facebook purchased Oculus VR, makers of the Oculus Rift, in 2014, for $2 billion.

"Bloody Vision" was used for the 1964 release of *The Incredibly Strange Creatures Who Stopped Living and Became Mixed-Up Zombies*, also known as *Teenage Psycho Meets Bloody Mary*.

In 2014, Google and other investors put over $500 million into the start-up Magic Leap, which seeks to make HMDs that project a digital light field—presumably a whole new world—into the user's eye.

INVENTING A NATION
AMERICA

No one person invents a nation, of course. The American project is one that still goes on today, and as it continues we involve more and more people in the work. It may never end, as we strive to get it right, and that will be just as well.

We are, more than almost any other country, one invented by the conscious work of man. We are the first nation to exist wholly in the modern world—a republic of laws, where everyone has a say, and anyone can join as long as they uphold those laws. We made ourselves, and we are constantly remaking ourselves as our world changes and we debate, over and over again, just what it means to be an American.

It hasn't been an easy process. There were plenty of missteps and flaws in the design. Our original sin of slavery, the false and pernicious doctrines of race. Our mistreatment of the peoples we found here, the land grabs we perpetrated against our neighbors. Like too many inventors, at times we have despoiled what we had in our rush to get ahead, get rich.

Yet when we've made it work, ours has been as free and prosperous and diverse a society as has ever existed on this earth—one that, uniquely, has been able to bring in masses of people from everywhere else to join in its success. Countless individuals, great and obscure, male and female, free and enslaved, of all colors and faiths and backgrounds, have contributed to that accomplishment.

Yet if one had to name a single founder who had the greatest vision of what America could be—and who, at the same time, best personified the American experience—it would have to be Alexander Hamilton.

Hamilton is having a certain vogue at the moment, right down to being the subject of a smash hip-hop Broadway musical. It's easy to see why. No other Founding Father was so constantly reviled, even in the raucous politics of his day. None had his ancestry or his loyalty to his country so constantly impugned.

Hamilton did not hail from an old family or the landed aristocracy. He was an immigrant, one who grew up in wretched and sometimes horrible circumstances in the West Indies. Abandoned by his father at a young age, orphaned by the early death of his mother—in the bed beside him, as both convulsed with fever—frequently "accused" of being of mixed race, or dubious religion, or a bastard, and lacking most formal education, he rose by sheer dint of his ability and ambition. His talents were so obvious to the local businessmen he clerked for that they sent him to study at

New York's King's College (later Columbia) at the age of seventeen. Just twenty-one (or possibly nineteen) when the American Revolution broke out, he distinguished himself as Washington's beloved personal aide and led a key bayonet charge that broke the British lines at Yorktown.

Yet even through the hardest days of the war, Hamilton kept his eye fixed on what sort of country he was fighting to create, studying government and economics whenever he could. Convinced that the Articles of Confederation could not hold the nation together, he wrote the resolution to hold a constitutional convention, then became, with James Madison, one of the two leading architects and defenders of the Constitution that resulted.

Appointed the nation's first secretary of the Treasury, he more than any other man made the literary construction that was the United States into a palpable reality. He built our first financial system, linking the nation together economically, making the states' separate debts into the obligation of all. He advocated always a national government that would be strong enough to lead, to enforce the law, to build a public infrastructure, to defend itself in the greater world—while still acquiescing to the will of the people.

None of this went off without a hitch or worked perfectly from the start—or ever since, for that matter. Hamilton's efforts brought us, however inadvertently, our first financial meltdown, first mistreatment of our veterans, first unfair and intrusive tax, and first use of American troops to subdue a popular rebellion. These blunders and excesses were in no small part the fault of the man himself—his zeal, his impetuosity, his passion, his hotheadedness, and his sometimes overwrought sense of honor. Such flaws would ultimately embroil him in what was also our national government's first sex scandal, and even get him shot dead, at much too early an age.

Yet if being all too much like the rest of us—like all of us—is what makes Hamilton beloved today, it's not what makes him great. Hamilton is great because he made America a country that could make things.

Much more than any other Founding Father, Hamilton understood that this was the nature of the modern world. Thomas Jefferson, for all his own vital contributions to our liberty, envisioned for America no more than a bucolic utopia of yeoman farmers. A more democratic version, in other words, of how mankind had always existed, tied to the land, churning out its bounty with the seasons.

Hamilton saw that in the new world, man would produce things himself. That as such, the world had become all about change. That all would have to take part, and there was no place in it for slavery, or other oppressive, feudal vestiges. That we would require education, capital, machines, cities, better communications, faster transportation, ways to bring greater and greater numbers of people into our common enterprise.

This was at the heart of inventing an America that would make possible so much more invention. It is the essence of us still. And even as we struggle with the messiness and sometimes the ugliness of democracy today, we must know that it is entirely within our power, our abilities, to make of America what we will.

Acknowledgments

America the Ingenious was a project written in near-record time, which would not have been possible without my wife and researcher, Ellen Abrams. El did amazing work in digging out all sorts of information about everything, even as the subjects and the focus of the book changed in the writing of it. She was tireless, uncomplaining, a constant source of encouragement . . . and my IT team, to boot. She has, as usual, all my love and thanks.

The idea for *America the Ingenious* was conceived by my publisher at Artisan, Lia Ronnen, who was just a delight to work with. My editor, Shoshana Gutmajer, was absolutely the best: unstintingly generous in her praise, her patience, and her insights. She was rigorous and perceptive, without ever losing her good temper and great sense of humor. Mura Dominko, assistant editor on the book, was also a joy to work with, smart and diligent, and always ready to jump into the breach. The rest of the Artisan team, including Zach Greenwald, Elise Nelson, Michelle Ishay-Cohen, Renata Di Biase, Allison McGeehon, Theresa Collier, and Nancy Murray, were all terrific and made a challenging project great good fun. They were meticulous in their work and their professionalism.

I would also like to thank my agent, Henry Dunow, for all his good work and my esteemed colleagues Jack Hitt, Daniel Okrent, Sam Roberts, and Brenda Wineapple for their generous praise. I am, as always, grateful for all the terrific people I am lucky to have in my life—my friends and family—and also for that tuxedoed, superhuman creature, our cat, Maisie.

Bibliography

Much of *America the Ingenious* was written with the help of the countless articles, diagrams, illustrations, and book excerpts that we are lucky to enjoy instant access to on that great American invention, the Internet. There were also many old-fashioned books that proved invaluable. A partial list of them follows.

Ambrose, Stephen E. *Upton and the Army*. Baton Rouge: LSU Press, 1993.

Beecher, Catherine E. *A Treatise on Domestic Economy, for the Use of Young Ladies at Home, and at School*. Boston: T. H. Webb, 1842.

Bryson, Bill. *One Summer: America, 1927*. New York: Doubleday, 2013.

Burns, Ric, and James Sanders, with Lisa Ades. *New York: An Illustrated History*. New York: Alfred A. Knopf, 1999.

Callahan, North. *TVA: Bridge Over Troubled Water, a History of the Tennessee Valley Authority*. New York: A. S. Barnes, 1980.

Chernow, Ron. *Alexander Hamilton*. New York: Penguin Books, 2004.

Collins, Gail. *America's Women: 400 Years of Dolls, Drudges, Helpmates, and Heroines*. New York: William Morrow, 2003.

Culvahouse, Tim, ed. *The Tennessee Valley Authority: Design and Persuasion*. New York: Princeton Architectural Press, 2007.

Denson, Charles. *Coney Island: Lost and Found*. Berkeley, CA: Ten Speed Press, 2002.

DeVeaux, Scott, and Gary Giddins. *Jazz*. New York: W. W. Norton, 2009.

Diehl, Lorraine B. *The Late, Great Pennsylvania Station*. Rockville, MD: American Heritage, 1985.

Eisenstadt, Peter, and Laura-Eve Moss, eds. *The Encyclopedia of New York State*. Syracuse, NY: Syracuse University Press, 2005.

Evans, Harold, with Gail Buckland and David Lefer. *They Made America: From the Steam Engine to the Search Engine: Two Centuries of Innovators*. New York: Little, Brown, 2004.

Gertner, Jon. *The Idea Factory: Bell Labs and the Great Age of American Innovation*. New York: Penguin Books, 2013.

Hamalian, Linda. *The Cramoisy Queen: A Life of Caresse Crosby*. Carbondale: Southern Illinois University Press, 2005.

Hiltzik, Michael. *Colossus: The Turbulent, Thrilling Saga of the Building of the Hoover Dam*. New York: Free Press, 2010.

Hitt, Jack. *Bunch of Amateurs: A Search for the American Character*. New York: Crown, 2012.

Hood, Clifton. *722 Miles: The Building of the Subways and How They Transformed New York.* New York: Simon & Schuster, 1993.

Jackson, Kenneth T., ed. *The Encyclopedia of New York City.* New York: The New-York Historical Society / New Haven, CT: Yale University Press, 1995.

Kaplan, David A. *The Silicon Boys and Their Valley of Dreams.* New York: William Morrow, 1999.

Kasson, John F. *Amusing the Million: Coney Island at the Turn of the Century.* New York: Hill and Wang, 1978.

Kevles, Bettyann Holtzman. *Naked to the Bone: Medical Imaging in the Twentieth Century.* New Brunswick, NJ: Rutgers University Press, 1997.

Kisseloff, Jeff. *The Box: An Oral History of Television, 1929–1961.* New York: Viking Press, 1995.

Lacey, Robert. *Ford: The Men and the Machine.* New York: Little, Brown, 1986.

Macaulay, David. *The Way Things Work.* Boston: Houghton Mifflin, 1988.

Miller, Donald L. *City of the Century: The Epic of Chicago and the Making of America.* New York: Simon & Schuster, 1996.

———. *Supreme City: How Jazz Age Manhattan Gave Birth to Modern America.* New York: Simon & Schuster, 2014.

Most, Doug. *The Race Underground: Boston, New York, and the Incredible Rivalry That Built America's First Subway.* New York: St. Martin's Press, 2014.

Nasaw, David. *Andrew Carnegie.* New York: Penguin Books, 2006.

———. *Going Out: The Rise and Fall of Public Amusements.* New York: Basic Books, 1993.

Reidenbaugh, Lowell. *Take Me Out to the Ballpark.* St. Louis: Sporting News, 1983.

Roberts, Sam. *Grand Central: How a Train Station Transformed America.* New York: Grand Central, 2013.

Rosenblum, Constance. *Boulevard of Dreams: Heady Times, Heartbreak, and Hope Along the Grand Concourse in the Bronx.* New York: NYU Press, 2009.

Snow, Richard. *Coney Island: A Postcard Journey to the City of Fire.* Brightwaters, NY: Brightwaters Press, 1989.

Starr, Kevin. *Golden Gate: The Life and Times of America's Greatest Bridge.* New York: Bloomsbury Press, 2010.

———. *Material Dreams: Southern California Through the 1920s.* New York: Oxford University Press, 1991.

Stevens, Joseph E. *Hoover Dam: An American Adventure.* Norman: University of Oklahoma Press, 1988.

Straus III, Roger, Hugh Van Dusen, and Ed Breslin. *America's Great Railroad Stations.* New York: Avery, 2011.

Thornley, Stewart. *Land of the Giants: New York's Polo Grounds.* Philadelphia: Temple University Press, 2000.

Ward, Geoffrey C., and Ken Burns. *Jazz: A History of America's Music.* New York: Alfred A. Knopf, 2000.

White, Richard. *Railroaded: The Transcontinentals and the Making of Modern America.* New York: W. W. Norton, 2011.

Williams, John Hoyt. *A Great and Shining Road: The Epic Story of the Transcontinental Railroad.* New York: Times Books, 1988.

Wilson, Richard Guy, Dianne H. Pilgrim, and Dickran Tashjian. *The Machine Age in America 1918–1941.* New York: Harry N. Abrams, 1986.

Wineapple, Brenda. *Ecstatic Nation: Confidence, Crisis, and Compromise, 1848–1877.* New York: Harper, 2013.

Yergin, Daniel. *The Prize: The Epic Quest for Oil, Money, and Power.* New York: Simon & Schuster, 1991.

Yungblut, Gibson. *Cincinnati Union Terminal: The Design and Construction of an Art Deco Masterpiece, vol. 1.* Edited by Linda C. Rose and Patrick Rose. Cincinnati: Cincinnati Railroad Club, 1999.

Index